MAN-POWERED AIRCRAFT

Other TAB books by the author:

No. 2205 *Aircraft Metalwork*
No. 2230 *Restoration of Antique & Classic Planes*

MAN-POWERED AIRCRAFT

BY DON DWIGGINS

MODERN AIRCRAFT SERIES

TAB BOOKS

Blue Ridge Summit, Pa. 17214

FIRST EDITION

FIRST PRINTING—FEBRUARY 1979

Printed in the United States of America

Library of Congress Cataloging in Publication Data

Dwiggins, Don.
Man-powered aircraft.

Includes index.
1. Human powered aircraft. I. Title.
TL769.D94 629.133'34 78-11137
ISBN 0-8306-9851-5
ISBN 0-8306-2254-3 pbk.

Foreword

The story of man-powered flight has held the interest of the author since the 1930s, when he set out to chronicle the history of flight in the United States prior to the Wright Brothers' success at Kitty Hawk in 1903. Considerable correspondence was held with many inventors long since gone, and much of this material, never before in print, appears here for the first time.

The 1977 success of the *Gossamer Condor* in capturing the £50,000 Kremer Prize for the first MPA flight around a figure-8 course does not, in the author's view, mark the end of the ages-old quest for controlled human powered flight.

Rather it stands as a challenge for the future, where many more astounding flights under muscular power are surely in store for adventurous souls who still long to fly with the freedom of the birds.

Don Dwiggins

Acknowledgements

The author is deeply indebted to many persons who over the years have contributed unpublished stories and anecdotes of their experimental efforts to achieve man-powered flight.

Special recognition must also be given to the curators of several museums, and librarians, who assisted in researching this history of human flying.

Among these are: The Smithsonian Institution's National Air and Space Museum, the Library of Congress, the Royal Aeronautical Society, and the Aviation History Library Committee of Northrop Institute of Technology.

Among the published works consulted were: Astra Castra, by Hatton Turnor, Chapman & Hall, London, 1865; The History of Aeronautics in Great Britain, by J. E. Hodgson, Oxford University Press, London, 1924; Histoire de L'Aeronautique, Charles Dollfus & Henri Bouche, L'Illustration, Paris, 1932; Artificial and Natural Flight, Sir Hiram Maxim, Whittaker & Co., New York, 1908; Through the Air, by John Wise, To-Day Printing & Publishing Co., Philadelphia, 1873; Progress in Flying Machines, by Octave Chanute, The American Engineer & Railroad Journal, New York, 1894; Travels in Space, by E. Seton Valentine & F. L. Tomlinson, Hurst & Blackett, Ltd., London, 1902; Man Powered Flight, by Dr. Keith Sherwin, Model & Allied Publications, Ltd., Herts., UK, 1971; and The History of Man-Powered Flight, by D. E. Reay, Pergamon Press, Elmsford, N. Y., 1977.

Several fine biographies of Leonardo da Vinci are available, including: The World of Leonardo da Vinci, by Ivor B. Hart, Viking Press, N. Y., 1961; The Mind of Leonardo da Vinci, by Edward MacCurdy, Jonathan Cape, London, 1928; and da Vinci's own Codice Atlantico, in the Biblioteca Ambrosiana at Milan, Italy, and a dozen da Vinci notebooks identified as A to M, in the Institut de France.

Contents

Chapter 1
The Gossamer Condor

The huge hangar at the edge of Shafter-Kern County Airport is silhouetted against the skylight of a false dawn. Toward the east, the start of a new day outlines the black contours of the Sierra Nevada, California's major mountain range, behind the lights of Bakersfield. Inside the hangar a monster sleeps.

This is the lair of the *Gossamer Condor*, an unbelievably grotesque giant flying machine held together with a maze of piano wire and aluminum tubing, a chimera, unlike any other aircraft in the whole world. It has absolutely no practical value, except for one thing—its entire raison d'etre is to win a £50,000 prize, the coveted Kremer International Competition Award, posted by the Royal Aeronautical Society of Great Britain.

Since ancient times, man has dreamed of flying under his own power, the ultimate freedom. In recent years human flight actually has been achieved a number of times, in different parts of the world. But such flights were merely extended straight-line ventures into the air.

To *really* fly, to be one with the birds, a man must be able to bank and turn, climb and glide, maneuver at will through the three dimensions of the sky. A few machines have in fact made right-angled turns, but none so far had the inherent ability to fly through a figure-8 course.

Dr. Paul MacCready (r) with son Tyler.

This was the whole idea behind the challenge of Henry Kremer, a British research engineer, whose hobby was developing the muscles of his body to lift weights. Could man's muscles really lift himself into the sky, and propel him around a figure-8 course, between two loci set half a mile apart?

Could he design a machine that could cross a ten-foot hurdle at the start and finish of the course? If he could, Kremer mused, then ancient dreamers like Leonardo da Vinci could be proven right. Man *could* fly like the birds!

Bryan Allen arrived at the big hangar from his home in nearby Bakersfield feeling optimistic. This would be the day! The day he would make history by being the first man in the world to complete the Kremer course. He hoped. He was uniquely qualified for the challenge before him.

At 25, Bryan had that unusual blend of genetic, athletic, and aeronautical characteristics required for the task. At 6 feet and 137 pounds, he also possessed the strength/weight ratio necessary for human-powered flight. For the past four months he had been training for this day.

Besides making numerous practice flights in *Gossamer Condor*, he had shaped up physically by bike riding and exercising on a device

Tyler MacCready checks controls of Gossamer Condor.

called an Ergometer. On this device, loads are applied to typical bicycle pedals and crank, and dials tell what speed and horsepower output the rider is achieving. Bryan learned he could sustain .35 hp for 30 minutes, .45 hp for seven minutes, and 1.2 hp for short bursts.

Gossamer Condor was designed with such a specific horsepower output-per-unit-time in mind, to make the man-machine relationship successful in meeting the demands of the Kremer course.

A special physical fitness program had already been set up for the project by Dr. Joseph Mastropaolo of Long Beach State University, at the request of Dr. Paul MacCready, designer of the *Gossamer Condor*. Months before, Dr. MacCready and his son Tyler, 14, had embarked on a rigorous exercise program, to shape them up for test flying the *Gossamer Condor*. In only six weeks, Dr. MacCready, at 51, developed the stamina of an average good college athlete.

Bryan Allen had another qualification for the task—besides winning his high school's annual Bicycle Day track race three years in succession, he also was an experienced hang glider pilot. While no special piloting skills were found necessary to fly the *Gossamer Condor*, Bryan already was familiar with the sensations of ultralight flight.

Dr. MacCready, too, had a background in flying, that was essential to his grasp of the requirements of man-powered flight. As a youth he built model airplanes and won his private pilot license at 16. As a glider pilot he became International Champion at a soaring meet in France in 1956, the first American to achieve that goal.

In addition, he earned his Master's Degree in Physics at CalTech in 1948 and his Ph. D. in Aeronautics at the same institution in 1952. Professionally, Dr. MacCready is president of a company he formed in 1970—AeroVironment, Inc.—involved in such areas as aircraft wake vortex investigations for flight safety, and wind power systems development.

The idea for building a machine to compete for the Kremer prize came to Dr. MacCready in July 1976, as he was working on an article comparing the flight of hawks to the flight of hang gliders. He saw that adapting hang glider construction techniques would permit construction of a very light, large, slow-moving vehicle. It would be easy to build, modify, and repair, and would require very little power to stay aloft.

First flight model of Gossamer Condor at Mojave.

Single-surface wing of first Gossamer Condor.

Within a few weeks he had built crude models of his proposed MPA, and then called in his close friend and vice-president of AeroVironment, Dr. Peter B. S. Lissaman, a South African by birth, who headed the firm's research work on wind energy programs for the government.

Dr. Lissaman had at his disposal a Hewlett-Packard 9821A computer, with which he was able to compress years of aerodynamic research and months of tedious computations into almost instant readouts. His goal was to develop critical airfoils for the main wing, the canard, and the propeller. The first two would be operating at very low airspeeds and at Reynolds Numbers between 500,000 and 1,000,000. Even the NASA wind tunnel people had left that region virtually virgin territory.

By October 1976, the *Gossamer Condor* was ready for rollout from its birthplace—the giant pavilion in Pasadena where the Tournament of Roses floats were constructed. Jack Lambie, a veteran soaring pilot and hang glider designer, helped to build the big machine, which spanned 88 feet and weighed a mere 50 pounds.

The building had to be vacated the next day, so the *Gossamer Condor* was rolled out at midnight of October 9, after only ten days' construction time. Curious visitors wondered what they were up to, but Dr. MacCready wanted the project kept absolutely secret, for good reason. The utter simplicity of the design might easily be copied and a rival machine might win the £50,000 Kremer prize ahead of the *Gossamer Condor*. In U. S. Dollars that was a tidy sum—about $87,500!

When one visitor asked Lambie what the darned thing was, he shrugged and replied: "Why, it's a new parade float, depicting the theme of government meddling in private affairs!"

There was a good reason for making the *Gossamer Condor* a crude machine—it could be rebuilt easily and quickly after the inevitable crashes it would endure, rather than taking months or a year to reassemble as the more sophisticated MPA's required.

The first version built in Pasadena, to check structure and pitch stability, had a wingspan of 88 feet, built around a single spar of 2-inch aluminum tubing. A maze of piano wires supported the wing from the top of an aluminum kingpost. Drag penalty of the wires, roughly 10% of the total parasitic drag, was negligible at the slow speeds the huge wing would be flying.

The wingspan and chord were determined by standard aerodynamic theory, while subsequent structural and aerodynamic design was aimed at permitting this wing plus a pilot to fly with a minimum of additional material. The initial criteria was limited to reliability and low power; the ability to maneuver in turns would come through trial and error.

Systems engineering and project management, often rather informal, were keys to the program's ultimate success. The overall approach was to try a simple solution for each problem encountered, thus more attention could be devoted to concepts that didn't work.

Eventually a new home was found for the project at Mojave Airport north of Los Angeles, where a pilot friend, Sean Roberts, had a huge hangar available. There the first operational prototype was built, with a bigger span of 95 feet, a chord of 11-¾ feet, and no streamlining of the pilot area.

The basic design concept was (a) support the wing spar with piano wire from the ends of a bottom post, king post, keel tube, and bowsprit, (b) have the pilot near the bottom of the bottom post to

Paul MacCready (on ladder) checks propeller blade of Gossamer Condor.

assist stability and pilot entry, (c) use the supine position to provide good forward visibility, to free the pilot's hands for controls, and to allow the main sprockets to be set right alongside the bottom post to take the chain tension, (d) put the propeller behind the wing for clearance and to keep its slipstream off the flight surface, (e) use a single surface airfoil for simplicity, (f) leave the pilot exposed for the time, but allow for future streamlining, and (g) ignore lateral control until satisfactory low-power straight flight was achieved.

After several design changes, the single-spar wing got a second spar to relieve compression and distortion of the ribs. Spreaders and wire bracing atop the airfoil helped relieve this problem.

Lissaman designed a 12-½ foot propeller with his computer magic, and developed a second, wider-blade version to permit the thrust to be increased 3 or 4-fold briefly for climbing or other

Plastic drive chain and idler wheel turns prop shaft of Gossamer Condor.

Pilot seat of Gossamer Condor #1.

"emergency" maneuvers. The latter propeller was used throughout the rest of the program.

Dr. MacCready's son Tyler made the first filght with the single-surface machine the day after Christmas, 1976, and stayed aloft 45 seconds after an assisted start. Through the next three months it was flown numerous times, and found to be too crude to fly for more than 2-½ minutes, far short of the time required to complete the Kremer course. It also flew badly in the slightest turbulence.

It was found that at the slow flying speed, even the lightest vertical turbulence altered the angle of attack along the wing locally, so that at any moment only a small portion of the wing would be operating within its narrow drag bucket. Lift distribution also would be altered far from the elliptical pattern desired.

Turns were next to impossible, due to high yaw forces that required extra power to overcome simply to stay aloft. A new "pilot" was found—Gregg Miller, a champion-caliber bike racer, with no flying experience but a fine set of well-coordinated muscles. Gregg

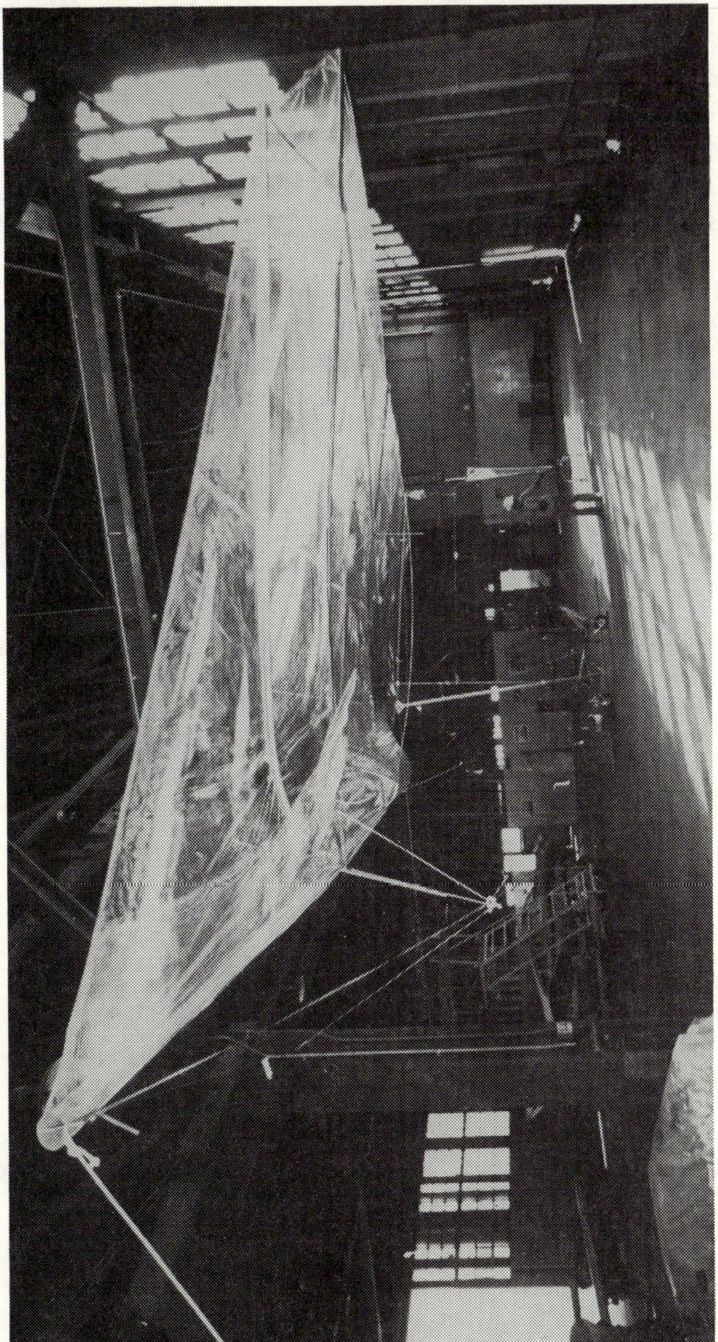

Gossamer Condor #1 at Mojave fills hangar.

could take off under his own power, and on his longest flight of 2-½ minutes covered more than 1,000 feet.

The author witnessed that amazing flight, which established a United States record for MPA's as it sailed serenely down the 10,000-foot runway of the ex-WWII Marine base at Mojave. Field elevation is 2,787 feet, but on that hot afternoon in January, the density altitude was maybe 4,000 feet.

Gregg Miller, a stocky 135-pounder, cranked slowly into the setting sun, some 10 feet off the deck, the red and yellow plastic propeller turning lazily at 110 rpm. He gripped the small steering wheel, turning it left and right to activate wingtip spoilers, and pushing and pulling to activate the canard out front for pitch control There was no rudder—do birds have rudders?

Paul MacCready sprinted alongside the machine, with his two sons, Tyler and Parker, both qualified MPA test pilots, and Jack Lambie. I could sense their overconfidence, and approved of it. Enthusiasm is needed on such a project.

There were even official observers on hand, in case Gregg did make it around the figure-8 course—M. W. (Woodie) Phillips, Technical Committee Chairman of the National Aeronautic Association, with two assistants, Alden DeWitt and Jess Bradshaw, stationed at the half-mile pylons. It was a bit ironic for these officials—only the summer before, they'd officially timed the world's fastest closed course flight of a USAF SR-71 bomber, at 2096 mph!

Gregg's legs finally tired, the *Gossamer Condor* settled down onto the runway, and Dr. MacCready knew the Kremer Prize was still unattainable. For a while, at least. A redesign was in order, and a new site had to be found, one with a much lower density altitude, and free from the unpredictable desert winds that lashed across the Mojave.

In March, 1977, operations were shifted to Shafter-Kern County Airport, in California's San Joaquin Valley. Work was immediately begun on a second flying *Gossamer Condor*, this one with a span of 96 feet, height 19-½ feet, length 30 feet, weight 30 pounds. The wing was double-surfaced, covered with ½ mil Mylar on top and ¼ mil Mylar underneath.

Design work now concentrated on stability and controllability, and a computer study revealed something interesting. Direct alteration of lift would have only a weak effect on roll, and a decrease in apparent mass at the wingtips was necessary, along with a decrease

Bike racer Gregg Miller flies Gossamer Condor #1.

in roll damping and considerable reliance on some yaw-roll (and roll rate) coupling.

A crude 36-inch balsa model was built, and pushed by hand through a swimming pool to study fluid accelerations, particularly those associated with the concept of apparent mass. The water tests confirmed results of the computer analysis, and so the wingtip chord was reduced to 5 feet, leaving the root chord at 10 feet.

This reduction in total wing area dictated that the pilot area be streamlined, and an 8-foot sweepback was added to meet balance and structural needs and help with the dihedral effect. The wing spar was moved back from the leading edge to the 28-½% chord position, and the ribs were made to cantilever on the spar and to hold their shape under load.

The double-surface airfoil was developed by Dr. Lissaman, using an interactive computer-human technique he developed for tailoring airfoils to special tasks. What he wanted was an airfoil of 11% thickness, nearly flat on the bottom, with a small pitching moment coefficient, high maximum C_l, low C_d over a wide C_l range, and sufficiently non-critical to cover mistakes in construction. Reynolds Numbers were between 500,000 and 1,000,000 in the concept.

In its final configuration, the *Gossamer Condor* wing used what was called the Lissaman 7769 airfoil, and its efficiency proved to be better than expected. The wing flew well and in combination with a rocking stabilizer, to provide a yawing force, it began showing good turning capability.

The final propeller, built by Lambie from Dr. Lissaman's computer figures, was made with balsa ribs and leading and trailing edges, set on a 1″ OD, 0.020″ wall tube, and covered with Monocote. It turned at 120 rpm, geared up from 90 rpm at the pedals.

By now Gregg Miller had left the project to go bike racing in Europe, and Bryan Allen took his place. On an early test hop, inadequate controllability resulted in an accident, and after rebuilding the wing it was found that false ribs were necessary to hold the proper shape. A fatter wing of 13.7% thickness was tried and rejected; it developed too much drag.

At this point Dr. MacCready knew he faced the most serious problem of all—how to make the *Gossamer Condor* turn properly. It was a problem that had stumped all other competitors in the Kremer

Gregg Miller flying Gossamer Condor #1.

The *Gossamer Condor* wing was more efficient than expected and showed turning capability.

competition. Initially, small ailerons were added to the trailing edge of the canard stabilizer in front, so that they could bank that surface, to lead the wing into a turn.

But what happened was, once the craft banked, instead of following through in an equilibrium turn, it just flew straight ahead, one wing low. "If you turned to the left," Dr. MacCready explained, "the left wing stayed low, but the nose pointed a bit to the right and you sort of sideslipped ahead."

He'd hoped that with the stabilizer putting in some side force, there would be sufficient dihedral effect to keep both wings balanced, with one wing traveling slower than the other, but it didn't work that way.

"Then, one Saturday I spent practically a whole day with my pocket computer, calculating what wing angles there should be in an equilibrium turn," he said. "It was an exercise where you note the airspeed differences between the inner and outer wings and figure how much the lift coefficient should be increased on the inner wing versus the outer wing. So, we put in the proper amount of twist, and it seemed to work okay!"

It was an historic breakthrough—never before had the problem of turning a low-speed aircraft of the MPA variety been licked. "The wing twist also maintained wing drag fairly constant," Dr. MacCready explained further, "and yielded a well-coordinated turn which could be trimmed with the canard roll. It turned out that the wing twist control, suitable for an equilibrium left turn, also served to initiate the left turn."

It was just the opposite of what you'd expect from adverse yaw on a lowered aileron, which in a conventional aircraft is neutralized with opposite rudder. In the *Gossamer Condor*, adverse yaw worked in the direction desired (left aileron or wingtip, down for a left turn). Instead of causing the left wing to rise in a bank to the right, the adverse yaw in reality swung the vehicle to the left, and yaw-roll coupling rolled the wing left, in a turn radius of about 250 feet, with a 3-degree bank angle.

Dr. MacCready now felt sure the Kremer prize was within grasp, but competition was becoming critical—in this country, two entries were making fast progress; Joe Zinno, in Rhode Island, was pedaling toward victory in his *Olympian* ZB-1, and Taras Kiceniuk, a CalTech student, was successfully testing his *Icarus* ground-effect

Ailerons on the stabilizer banked the craft, but the craft wouldn't turn.

MPA. Meanwhile in Japan, Professor Hidemasa Kimura's balsa-wood *Stork* MPA has covered more than 6,000 feet!

The *Gossamer Condor* had made more than 400 flights, and frequently had been wrecked, six times at Mojave and three times at Shafter. Sometimes the wing snapped, other times the fuselage

Gregg Miller warms up on bike prior to flight.

Dr. Peter Lissaman designed wing of Gossamer Condor on computer.

collapsed. But Dr. MacCready had expected that, and each time she was rebuilt, better than before.

Finally, comes the day when both man and machine are ready for the big one. Bryan Allen takes his place inside the Mylar cocoon, and warms up on the pedals in the light of the predawn glow. The official observers are in position. The whole crew is tense, waiting.

Standing beside the *Gossamer Condor*, watching, are Dr. MacCready, his two sons, Jim Burke, Vern Oldershaw, the main constructor, and others. The propeller turns slowly, driven by a plastic cable-chain linking two aluminum wheels, one of 62 teeth, the other of 52 teeth. Its translucent red and yellow blades become a blur. The machine moves forward, slowly, on its two tiny plastic wheels.

Suddenly the *Gossamer Condor* is flying, climbing slowly toward the starting point, midway between the two pylons, half a mile

Tyler MacCready flies Gossamer Condor #1.

apart. Lambie is waiting there now, holding the ten-foot hurdle Bryan must pass over. Lambie watches intently, counting Bryan's pedal strokes...16 strokes every ten seconds, roughly 98 rpm.

"Come on, Bryan!" someone yells. "You'll make it!"

Rhythmically and gracefully, Bryan Allen pedals on toward the first pylon. He easily banks his wing, sweeps around it, smoothly. Back on the straightaway, he heads for the second pylon, as the noise of an approaching Ag plane disturbs the early morning stillness in the distance.

"You're doing great!" a voice cries. "Don't stop now!"

The *Gossamer Condor* approaches the second pylon, banking in the opposite direction to circle it, and fly back across the finish line at the midway point. A wing goes down, then suddenly it shudders. The stabilizer pitches up, the wing snaps, and folds back.

"Oh, no!" someone cries. Everybody rushes forward to catch the slowly collapsing machine. It sinks to the ground, a broken bird. Dr. MacCready's heart sinks too. With triumph so close, what happened?

Ironically, a trailing vortex from the Ag plane, landing half a mile away, has snapped the wing like a match-stick! The very thing his firm, AeroVironment, has spent so much effort studying!

Glumly, they carry the *Gossamer Condor* back to the huge hangar, for one more reconstruction. But is there time left? Dr. MacCready's mind turns back, across the centuries, to the long distant past, when man first dreamed of accomplishing what he was trying to accomplish—controlled flight on the power of the human muscle alone. To the time of Leonardo da Vinci, nearly five centuries ago...

Chapter 2
Da Vinci's Great Bird

Leonardo da Vinci stood gazing at a V-formation of soaring birds floating effortlessly in the sky above Monte Ceceri, near his villa at Fiesole, in the environs of Florence in northern Italy. The lovely, green hill, 1358 feet high, was shaped like the body of a swan (Monte Ceceri means "Swan Mountain"). His keen eyes studied how skill-fully the birds maneuvered their wings and tail feathers to catch the slope winds, instead of beating furiously against the air.

Soaring birds simply let the winds do their work for them, "like the bird of prey, the vulture, which I saw on my way to Fiesole, on March 14, 1505," he wrote later. Here, he was certain, lay the secret of man flight. He had been all wrong in the past. For sixteen years da Vinci had dreamed of flying, and in those years, from 1483 to 1499, he had secretly invented and sketched no less than 14 manpowered aircraft to be propelled by flapping wings.

The study of birds had, in fact, been a lifelong obsession for da Vinci. During his turbulent years as a famous painter, he found relaxation sketching lovely mechanical drawings of ornithopers, helicopters, and parachutes. Any device he could think of to escape to the freedom of the sky.

It was an urge that had gripped men before him, but da Vinci was first to get it all down on paper, in detailed sketches. He described these flying machines carefully in his secret code—da Vinci was left-handed, and wrote from right to left, in a tight scrawl.

People today call it mirror writing, because it appears normal when viewed in a mirror.

Many of his drawings have been lost, but today two prime sources are filled with his fascinating concepts of mechanical birds. One is a collection of his work known as *Manuscript B*, in the Institut de France, which he started in 1488 at the age of 36. The other is his famous *Codex Atlanticus*, in the Ambrosiana Library in Milan.

In 1505, the year he turned from mechanical wing-flappers to the study of soaring birds and the unknown science of meteorology, da Vinci compiled a small notebook of some thirty pages, now in the Royal Library in Turin. In this notebook, called the *Codice Sul Volo degli Uccelli* (Manuscript on the Flight of Birds), da Vinci reveals how close he came to grasping the true secret of flight.

In his *Codex Atlanticus* he had written: "A bird is an instrument working according to mathematical law, an instrument which is within the capacity of man to reproduce with all its movements, though not with a corresponding degree of strength, for it is deficient in the power of maintaining equilibrium. We may therefore say that such an instrument constructed by man is lacking in nothing except the life of the bird, and this life must needs be supplied from that of man."

Later on da Vinci would write: "To attain to the true science of the movement of birds in the air, it is necessary to give first the science of the winds, which we will establish by means of the movements of water."

Da Vinci had long been fascinated by hydrodynamics, and marveled at the similarity of the flow of water down-stream around rocks and that of air over hills and valleys. He studied the action of wind on sails, and how a breeze could draw a sailboat forward, into the wind.

Some 400 years later, Professor Samuel P. Langley of the Smithsonian Institution would write a book on *The Internal Work of the Wind*, one of the earliest studies of the kinetic energy of the wind.

But in the 15th Century, da Vinci had no predecessor to study, only his own keen powers of observation, with his artist's eyes. This dependency on originality in his investigations has been cited as the main reason da Vinci "never got off the ground." Waldemar Kaempffert, science editor of the *New York Times*, wrote in the 1961 edition of the prestigious *Encyclopedia Britannica:* "It was

Self portrait by Leonardo da Vinci.

largely lack of a technical heritage that made it impossible for Leonardo da Vinci to invent a practical flying machine."

Da Vinci, of course didn't know he was up against the impossible. He clung to his great dream of flying, and wrote in his notebook: "The great bird will make its first flight—upon the back of its great swan—filling the whole world with amazement, filling all records with its fame, and bringing eternal glory to its birthplace."

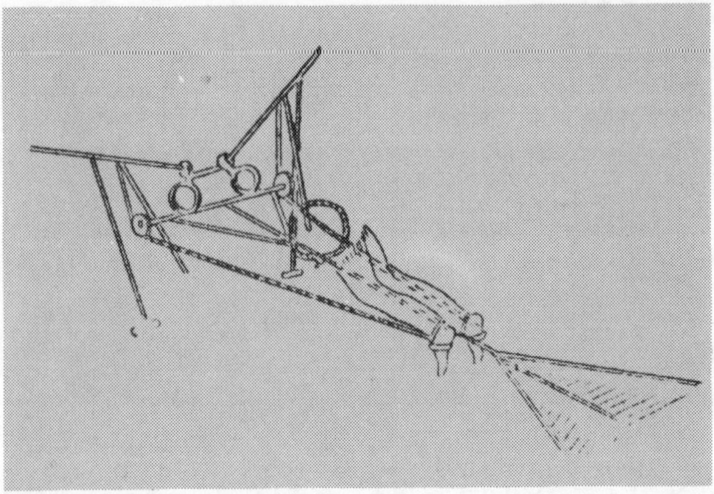

Sketch shows how da Vinci MPA pilot would fly prone with feet in stirrups to flap wings.

No record exists that da Vinci ever succeeded in flying from the top of Swan Mountain, or elsewhere, but there is good evidence that he tried. In his *Codex Atlanticus* he wrote:

"Tomorrow morning, on the second of January 1496, I will make the leather for the straps and the trial." (At that time da Vinci had proposed making experimental flights from the roof of a house in the Corte Vecchia at Milan, where he could practice unobserved by workmen then engaged elsewhere, on the dome of the Cathedral.)

But it was the crest of Swan Mountain where da Vinci hoped to find glory. To him, painting and sculpture were simply expressions of his own talent. He believed he would be remembered rather as the first man to escape the bondage of gravity and soar off into the sky. That would be real immortality!

Da Vinci had reasons to become so introverted where his mechanical inventions were concerned. These machines were pure creations of his brilliant mind, the end result of powerful forces at work—keen observation, intellectual understanding, and a practical approach to solving seemingly insurmountable problems.

Other things had gone badly for him. A vast effort to divert the Arno River and isolate the rival town of Pisa from the sea ended in disaster, when spring flooding and a violent storm wrecked the project. Gascon invaders had vandalized his great equine colossus in Florence, by using it for archery target practice. His great mural of a

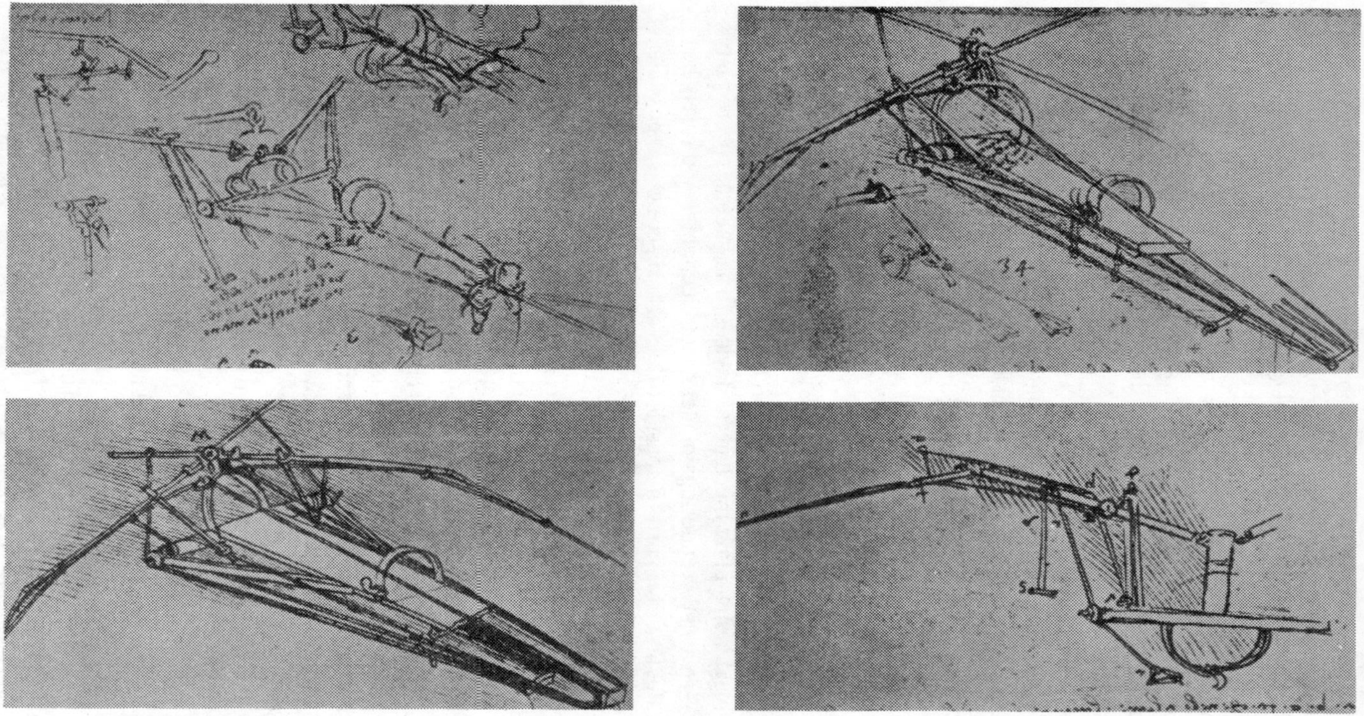
Various ornithopter MPA designs by da Vinci.

battle scene at the Sala de Consiglio was ruined, when the paint ran when he tried to speed its drying with a bonfire.

Younger artists in Florence were challenging his position and reputation. Michaelangelo publicly attacked him. Raphael unconscionably copied some of his greatest works. Of these, one—his beloved Madonna (Mona) Lisa—was the one painting that would establish his reputation through the ages as the greatest of the Florentines.

All this was nothing to da Vinci compared to his desire to fly. It was a desire misunderstood through the years, by "experts" seeking to understand his genius. A childhood fantasy, of a great vulture landing in his crib, was interpreted by Dr. Sigmund Freud as evidence of homosexuality. Charles H. Gibbs-Smith, the noted British aviation historian, wrote that in his opinion da Vinci's flying machine designs were simply too heavy to fly, and that he could not have invented the helicopter because spinning propeller toys already were common.

Da Vinci's personal life has no bearing on his achievements of the mind in inventing mechanical devices. His sketches were just that—an artist's renderings of ideas, not blueprints. And surely he was the first of record to consider the vertical screw as a possible means of lift for man-powered aircraft.

Many of his basic observations were far ahead of their time in the science of aerodynamics, which would not come to full flower until the beginnings of the 20th Century, with the 1905 statement of the Circulation Theory of Lift by Kutta and Zhukovski.

Until that time, and in fact since then, many observers pinpointed only a part of a wing's total circulation as responsible for lift. In both ornithopter and rigid wing aircraft designs, lift was assigned to a negative pressure area on top and a positive pressure area below the wing. The effect of vortex flow *behind* the wing was entirely overlooked.

Thus, da Vinci may be excused for not grasping the full picture of how lift is created by a wing moving through the air, although he was first to postulate the partial theory of positive and negative regions of air pressure below and above a bird's wing.

Perhaps he had in mind the analogy of a swimmer's hand in water, or that of a boat's oar, when he considered the behavior of air as a fluid. He actually preceded Sir Isaac Newton's law of action and reaction by a good 200 years—in his early studies of bird flight, da

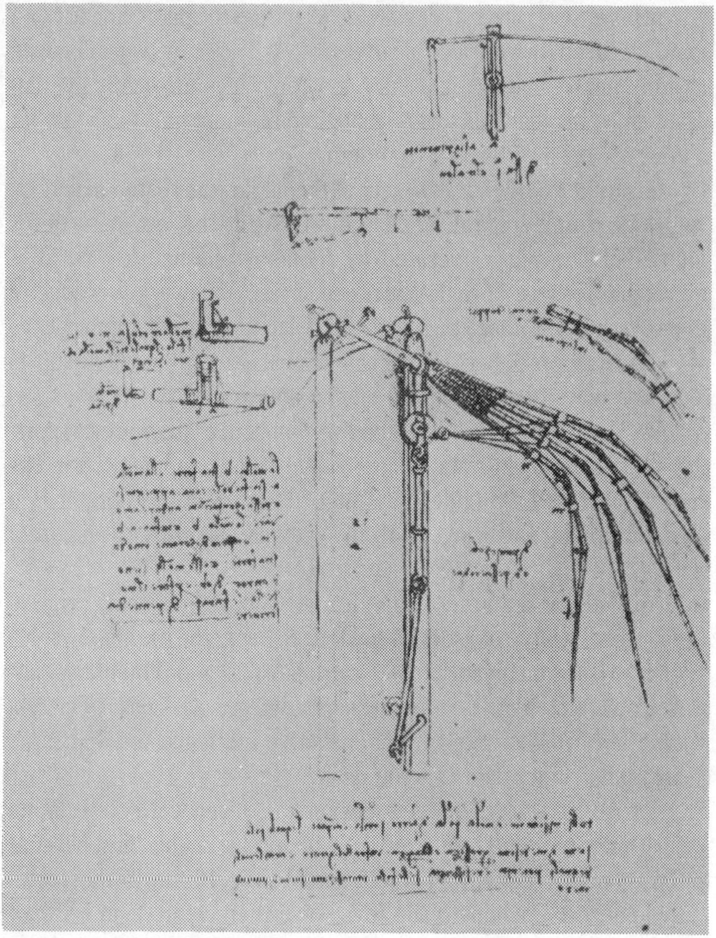

Da Vinci sketch shows mechanical linkage of MPA wing.

Vinci wrote in his Codex Atlanticus: "An object exerts the same force against the air as the air against the object. A man with sufficiently broad and properly constructed wings could learn to overcome the resistance of the air, and, by conquering it, subjugate it and rise above it."

To da Vinci, nature was the best teacher; to fly like a bird, one must have wings that flap, he considered early in his studies. Frequently he would go to the market place and purchase small birds from hawkers, then set them free. Birds did not belong in cages, and besides, he could study their flapping mechanism closely that way.

In his earliest aircraft design, da Vinci had the pilot lying flat on a board, secured by iron hoops. He operated the wings by pulling with his arms and treading with his more powerful legs. From his anatomical studies, he was well aware that a man's chest muscles were far inferior to those of birds in strength.

In another model, the weight of the apparatus rested mainly on the pilot's shoulders, rather than the board he lay on. It carried a double set of wings, operated by foot stirrups and levers. It is interesting to note how da Vinci used his extensive knowledge of leverage and systems of pulleys to increase the "thrust" of the wings when used as propellers, in a manner analagous to employment of reduction units today between engine and propeller.

Da Vinci came to feel that the flyer could operate more comfortably and efficiently in a standing position than when lying down. His third flying machine model thus had the pilot sitting upright in a gondola, driving the wings with foot pedals linked by a system of pulleys.

An unusual feature of this design was a retractable tricycle landing gear, which also served as steps to enter the machine. With a stretch of the imagination, one can also see a rudimentary "autopilot" in a sketch of a pendulum suspended in a glass ball. "This ball within the ring," he noted, "will enable you to guide the apparatus straight ahead or aslant, as you wish."

There quickly followed a design for a vertical screw lifting mechanism, frequently referred to by historians as the "first helicopter" design. A steel spring would actuate the 16-foot blades, with stored energy. Spinning blade toys were then common in Europe, based on imported Chinese spinning tops, but da Vinci seems to have been the first to suggest using a lifting screw for a man-carrying aircraft.

In 1496 da Vinci designed still another flying machine powered with coiled springs, which transmitted power to flap the wings through a system of reciprocating cranks and levers. With this device, da Vinci temporarily abandoned his researches into flying mechanisms, turning to other projects.

The last decade of the 15th Century was one of turmoil. The New World discoveries begun by Christopher Columbus were well under way, and in Florence French invaders of Charles VII held sway.

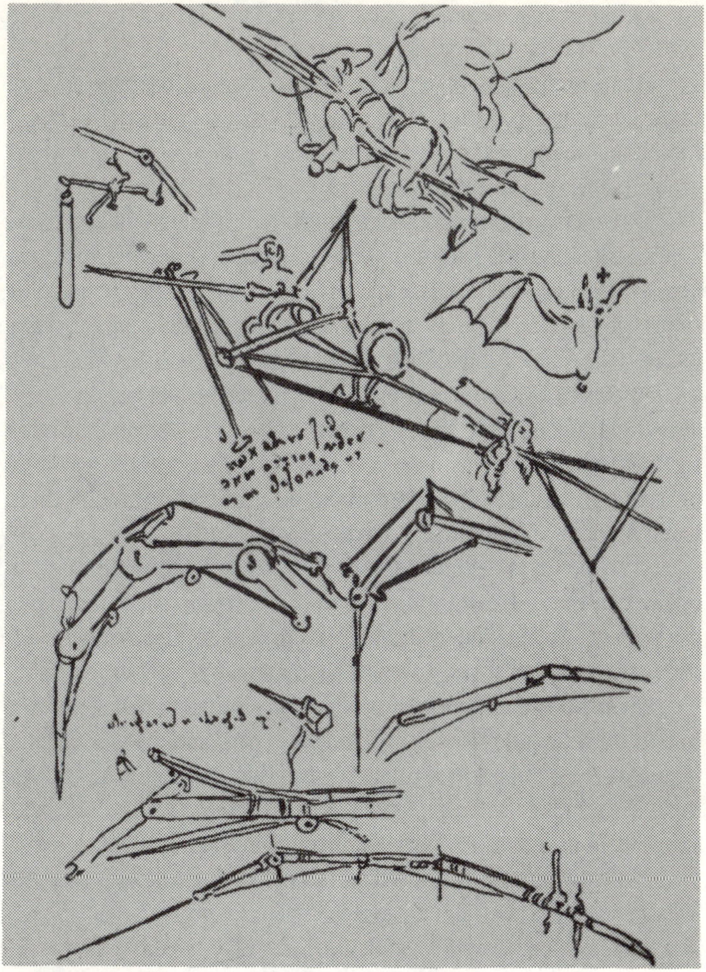

Da Vinci studied bat wing as model of MPA design.

Offering his services to Ludovico Sforza, the tyrant of Milan, da Vinci designed an amazing collection of military machines, including chariots with revolving knives on the axles. In a playful mood, he left instructions for the launching of stinkbombs to repel enemy invaders.

When Sforza's rule was overthrown, da Vinci returned to Florence, only to learn that another inventor, the mathematician Giovanni Battista Danti, had attempted to fly in a man powered aircraft, in January, 1503. Danti had jumped from the tower of the Church of Santa Maria della Vergine, but his wings hung up on a cornice and he fell to earth, breaking a leg. Da Vinci felt it was time to settle down to serious study of acronautics.

Abandoning his earlier work on mechanical flapping contraptions, he turned to a study of air currents blowing up and over Swan Mountain, and how the great birds flew for hours on silent wings, drawing energy from the winds.

Da Vinci's writings contain numerous statements of broad concepts he called the "four potentials of nature"—*weight*, *force*, *motion*, and *impact*. Here was a starting point, from which to solve the riddle of heavier-than-air flight—an understanding of the same forces responsible for flight we now call lift, gravity thrust, and drag.

At this time, da Vinci reexamined his designs of man-powered aircraft patterned after birds. There was one thing wrong, he felt—the feathers. There was no practical way to duplicate their amazing lightness and interlocking construction. Air leaked through them, when he built artificial wings covered with feathers.

Da Vinci drew sketches of wings constructed with slats that opened on the upstroke and closed on the downstroke, something like Venetian blinds. In later years, similar patented devices were tried, but failed to work, when others sought to imitate da Vinci.

Then he thought of the bat. In his notebook *Sul Volo degli Uccelli* he reminded himself: "You are to remember that your bird (flying machine) ought not to imitate anything but the bat, because the membranes form an armour, a strength to the wings. The bat is aided by the membrane, which binds the whole and is not pervious."

His batwings were to be built of cotton fustian cloth, starched silk, and a fiber mesh for the outer trailing edge surfaces. Ribbing was to be of cane. In appearance, the Great Bird of da Vinci was not unlike one of Otto Lilienthal's successful early monoplane gliders, of the 1890's.

Plunging further into basics, he formulated the law of inertia more than a century before Galileo, in 1638, claimed credit for it. Da Vinci wrote: "Every movement tends to maintain itself, or rather, every body in motion continues to move so long as the influence of the force of its motor is maintained in it."

Once launched from Swan Mountain, a glider should thus continue on its way, the same way a bird soars. But what of equilibrium and balance? He invented a device to determine a bird's center of gravity, and wrote: "When a bird sinks, the center of gravity is moved beyond the center of resistance, and when the bird wishes to

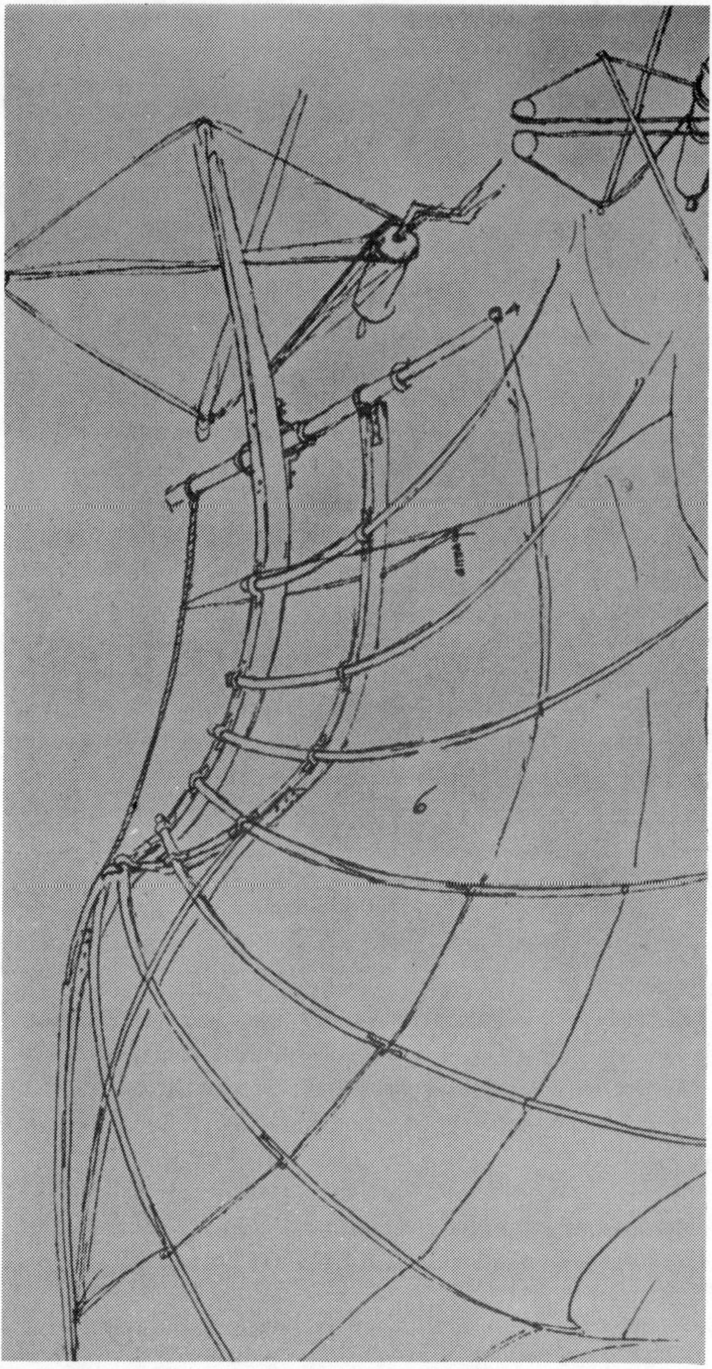

Da Vinci switched from birds to bats for MPA wing design.

rise, the center of gravity remains behind the center of its resistance."

In one of his later sketches, da Vinci included a cruxiform design for a tail with both horizontal and vertical surfaces, to be controlled by head movements. It was a forerunner of today's aircraft empennages.

In the spring of 1506, da Vinci's problems had reached a high point. He had quarreled with Pope Julius II, Michaelangelo, and others. His painting was going badly, particularly his great battle scene in the Sala del Consiglio, whose colors had run all together when the room was overheated.

From the villa of his adoptive uncle, Alessandro Amadori, at Fiesole, he gazed longingly at the summit of Swan Mountain. What an achievement it would be, to fly from there and soar with the birds!

What happened may never be known.

A legend has been reported by da Vinci's biographer, Antonnia Vallentin, that "one day there rose up from Monte Cecero a huge bird, as though Swan Mountain handlifted itself into the air. Suddenly the bird disappeared, as though swallowed up by the skies."

After da Vinci's death, Girolamo Cardano, son of Leonardo's close friend Fazio Cardona, would write: "Both those who have recently attempted to fly came to grief. Leonardo da Vinci also tried to fly, but he, to, failed. He was a magnificent painter."

Not until the 18th Century were da Vinci's amazing diaries made public, with their intricate notes and drawings of flying machines. According to Cecil Hilton Monk Gould, Assistant Keeper of the National Gallery in London, "It is on his drawings that Leonardo's claim to be considered one of the greatest of artists must chiefly rest..."

Certainly his brilliant mind and quick eye looked into the future, and saw the sky filled with flying machines. Whether or not he actually flew one remains unimportant. His contribution to aeronautics was originality of perception and design. By the same token, when the Wright Brothers first flew, four hundred years later, their contribution was not the invention of the airplane—da Vinci, Lilienthal, Moulliard, Chanute, Langley, and others already had invented it. The Wrights were the first to *learn how to fly*.

Chapter 3
The Birdmen

The origins of man-powered aircraft are lost in antiquity. In mythology, the best known legend of manned flight is the story of Daedalus and Icarus, with its counterpart in the Scandinavian legend of Wayland, the smith who forged his wings. Skipping over Bladud, the legendary Flying King of Britain, in the ninth century, and the apocryphal flight of the sorcerer Simon Magus, we come to Roger Bacon (1214-94), who dreamed of flying 200 years before Leonardo da Vinci.

Imprisoned for his radical teachings against beliefs of theologians and scholars of his day, Bacon's ideas on flying have frequently been misunderstood by aviation historians. In Bacon's time it was thought that the atmosphere, like the oceans, had an upper surface that could be sailed upon. Bacon envisioned an aerial machine as a "hollow globe of copper, or other suitable material, wrought extremely thin and filled with 'etherial air or liquid fire' and launched from a mountaintop."

"There may be some flying instrument," he wrote, "so that a man sitting in the middle of the instrument, and turning some mechanism, may put in motion some artificial wings which may beat upon the air like a bird flying."

According to the American balloonist and aeronautical historian, Professor John Wise, "Soon after Bacon's time projects were instituted to train up children from their infancy in the exercise of

Legendary flight of Icarus took him too near the sun, which melted wax in wings and caused crash.

flying with artificial wings, which seemed to have been the favorite plan of the flying philosophers and artists of that day."

We have seen how completely absorbed da Vinci was with the subject of imitating bird flight, but over the ensuing centuries the goal of constructing a practical ornithopter, or wing-flapper, has remained elusive.

Curiously, an attempt to cross the English Channel in a Man Powered Aircraft (MPA) seems to have occurred more than four centuries ago, in the fall of 1507, the year after da Vinci was contemplating a flight from the summit of Monte Ceceri.

John Lesley, in his *History of Scotland*, published in 1578, relates that an Italian-born adventurer, John Damian, leaped from the top of Stirling Castle with homemade wings of chicken feathers, hoping to overtake a boatload of ambassadors headed for France. "But he fell to the ground and brake his thee bane," Lesley related, ascribing the failure to "a natural affinity of the hens' feathers to return to the dunghill."

Two accounts of MPAs appear in the 17th Century. A French tightrope dancer named Allard reportedly tried to fly from the terrace at St. Germain before Louis XIV in 1660, but crash-landed. In 1678 another Frenchman, the locksmith Besnier, designed a pair of oscillating wings of muslin, with shutters that opened on the upstroke and closed on the downstroke, in the manner of da Vinci's device.

In his 1894 book, *Progress in Flying Machines*, the brilliant American civil engineer and aviation enthusiast Octave Chanute reported that Besnier actually sailed over the roof of an adjoining cottage, then later sold his wings to a traveling mountebank, who performed with them at fairs.

Eighteenth Century would-be birdmen continued the quest of man-powered flight in crude ornithopters, before invention of the hot-air balloon by the Montgolfier brothers in 1783 changed the whole picture.

The Marquis de Bacqueville in 1742 made a valiant try, hoping to flap his way across the Seine River in Paris, from a window of his mansion to the Tuilleries Gardens, some two hundred yards away. At mid-river he faltered and fell onto the desk of a washerwoman's barge, breaking a leg.

More fortunate was the Abbe Desforges, a canon of the church of Saint-Croix at Etampes, who tried to get airborne in a winged

Octave Chanute published history of MPA's in 1894.

chariot in 1772. The machine featured a pair of wings and a horizontal plane, with a total area of some 145 square feet. Four men assisted in the launch, but the harder he flapped the worse things became—the action of the wings, Chanute relates, pulled him down instead of up.

With the introduction of ballooning, inventors of wing-flapping devices could devote their full energies toward achieving forward propulsion rather than lift. Among the first was Jean-Pierre Blanchard, a 28-year-old Frenchman who in 1781 had attempted to fly a MPA fitted with four wings. He later admitted the wings merely "served to agitate an indocile element, with no more effect than those of a heavy ostrich."

In March, 1784, Blanchard adapted his "vaisseau volant" wings and rudder to a hot air balloon, history's first attempt to navigate the air in a lighter-than-aircraft. He discarded the devices as too heavy, but later successfully used a pair of silken oars and a revolving fan to tack upwind. Blanchard, incidentally, became the first airman to cross the English Channel, in 1785, and later introduced ballooning to America.

On August 23, 1783, some 50,000 Frenchmen gathered in the Champ-de-Mars in Paris to witness the first public balloon ascension in that city—a hydrogen sphere built by Anne-Jean and M. N. Roberts for the noted physicist Jacques Alexandre Cesar Charles. Among the witnesses was the United States Ambassador to France, Benjamin Franklin.

After the ascension, someone remarked to Franklin: "What good can a balloon be?"

Franklin's retort was the bon mot of the day: "Eh, a quoi bien l'enfant qui vient de naiture?" (What good is a new-born babe?)

If Franklin was skeptical of ballooning, his ingenious mind was fired with the concept of man powered flight. He sat down in his study and sketched an MPA design that historians have largely overlooked. It consisted of a flying machine with a huge umbrella of quill feathers to be flapped up and down by a rider suspended below, his feet in stirrups.

Like da Vinci before him and Dr. Paul MacCready nearly 200 years later, Franklin recognized that legpower would be the prime mover of the human engine in any sucessful MPA.

As to ballooning, he would write: "These machines must always be subject to be driven by the winds. Perhaps mechanic art may find easy means to give them progressive motion in a calm, and to slant them a little in the wind."

Dirigibles are another story, of course, but over the years aeronauts have sought to navigate the air with muscle power. In the 1880s, Carlotta Meyers, a lady balloonist of Mohawk, New York, used a folded newspaper to waft her way across the sky in a sport balloon, and at the beginning of this century, Roy Knabenshue used

American dirigible pilot Roy Knabenshue rowed his craft across the sky with silken oars in early 1900s, copying feat of 18th Century Italian balloonist Vincent Lunardi.

silken oars like those of Blanchard and Vincent Lunardi to "row" his dirigible across the skies of Pasadena, California.

Another aeronautical pioneer inspired by the invention of the balloon in 1783 was Sir George Cayley, remembered today as "the Father of Aeronautics" in Great Britain. Like da Vinci, he studied bird flight and came to the conclusion that wing flapping "was for the birds" and not for man. He summarily dismissed ornithopter flight as "ridiculous."

Cayley would win fame with a series of successful glider flights in a machine of some 300 square feet of surface, beginning in 1808. A story is told that his coachman flew the machine 300 yards across a valley and crashed.

Afterward, the coachman, in his broadest Yorkshire dialect, said, "Please, Sir George! I wish to give notice! I was hired to drive, not to fly!"

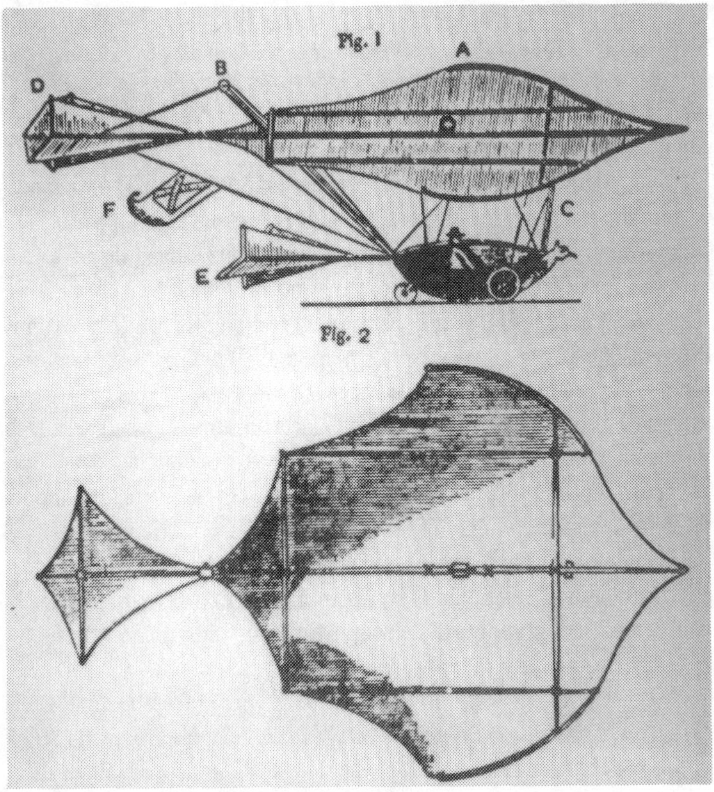

Early aircraft design of Sir George Cayley.

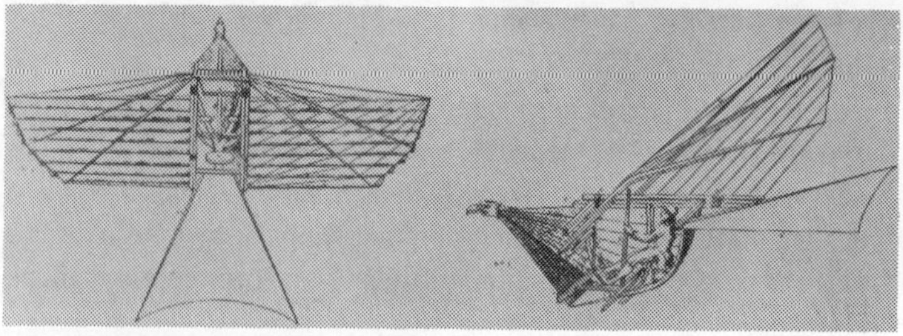

1810 MPA design of Thomas Walker.

A contemporary of Cayley, Thomas Walker, in 1810 published a work entitled *Treatise upon the Art of Flying by Mechanical Means*, with a subtitle: *Instructions and Plans for making a flying car with wings, in which a man may sit, and by working a small lever, cause himself to ascend and soar through the air with the facility of a bird.*"

Like da Vinci, Cayley and others, Walker studied bird flight and dissected many birds, studying "very minutely the mechanism of their wings, tails, and all parts which they employ in flying." The secret of bird flight apparently eluded Walker, however, as no record exists that he built such an MPA as he envisioned and wrote about.

The work of Cayley and Walker, nevertheless, had their impact on the history of flight. William Samuel Henson and John Stringfellow, two contemporary Britishers, were inspired by their ideas to build steam-powered model aircraft, one of which actually flew in 1848, 55 years before the first powered flights of the Wright Brothers at Kitty Hawk, North Carolina. Work on ornithopters continued with little or no success in Euope through the 19th Century, with the possible exception of Charles Spencer, who exhibited his machine at the first Exhibition of the Aeronautical Society of Great Britain in 1868. A 140-pound athlete, Spencer tucked under his arms a machine weighing only 24 pounds. Two articulated wings of 15 square feet each were attached to a rigid plane of 100 square feet, with a small tail behind. By sprinting down a small hill, he was able to flap along above the ground for distances up to 130 feet, probably despite the movable wingtips.

Something new was suggested in 1873 with publication of J. Bell Pettigrew's *Animal Locomotion, or Walking, Swimming, and Flying*, in which the author pointed out that birds in flapping flight flex their wings so that the tip feathers describe a figure-8 motion. In

1879 an experimeter named Dandrieux built an ornithopter in which the wings were attached to an oblique axle to attain the figure-8 motion, but nothing came of it.

For a time, the ornithopter idea was put aside in favor of other MPA devices, none of which worked. As early as 1808 Cayley had tried a set of umbrella-shaped wings unsuccessfully, and from 1809-12 Jakob Degen, an Austrian clockmaker, failed to fly in a similar device of taffeta wings that operated with a valvular action. He did rise to a height of 54 feet beneath a small balloon, but went nowhere.

In July, 1854, the French aeronaut Letur ascended beneath a balloon from London's Cremorne Gardens in a flying machine that combined a parachute and wings operated by a treadle. Adverse winds blew him into a forest near Tottingham, tumbling him to earth. He died ten days later.

Vincent de Groof, a Belgian shoemaker, was killed in a similar accident 20 years later at Cremorne Gardens, attempting to fly from a balloon launch in a contraption that was basically an ornithopter with wings of 20 square meters area. He flapped the wings with his arms and wiggled a tail with his feet, but during the fall the wings folded upward and collapsed. He was killed on the spot.

Further experimentation in MPAs during this early period involved the design of ornithopters with much larger wings. While more likely to sustain the pilot, they were simply too big to flap fast enough. An example was the MPA of C. F. Meerwein, architect to the Prince of Wales, whose calico wings measured 111 square feet of area, for a wing loading of roughly .55 lb/sq. ft., about that of a wild duck. It wouldn't budge.

Chanute reports that an inventor named Breant in 1854 attempted to find a solution to flapping big wings by using elastic cords to assist the upstroke, along with three valves. His wing area totaled 108 square feet. Nothing happened.

Jean-Marie Le Bris, a sea captain of Brittany, in 1857 got airborne in an unusual MPA patterned after the albatross, a graceful soaring seabird he had studied during voyages around Cape Horn. He had placed his machine on a horse-drawn cart at the top of a rise facing into the wind, on the coast of France near Brest. The wings spanned 23 feet and had a surface of 220 square feet, and their angle of incidence was controlled by pulleys and cords actuated by levers. The horse galloped downhill, the machine rose to 300 feet, at the end

Belgian shoemaker Vincent de Groof's MPA collapsed in launch from balloon. He was killed.

of a tether. Encouraged, Le Bris tried again in 1868, took off, crashed and broke a leg.

To Chanute, this failure was evidence that man would never fly under his own muscle power. He wrote in 1894: "There seems to be no hope that any amount of ingenuity or skill can enable man to accomplish this feat."

He concluded that the fixed-wing aeroplane held greatest promise for conquest of the air, and in fact could support nearly twice the

French sea captain Jean-Marie Le Bris patterned his MPA after the graceful albatross, once rose 300 feet.

weight per horsepower as the ornithopter. The big problem remaining, he believed, was the matter of equilibrium. If one intended to pattern an MPA after birds, he would do well to forget the flappers and study the soaring species. "All birds are acrobats," he wrote, "but the soaring kind, if closely observed in a gusty wind, will be seen to perform feats of balancing more delicate and wonderful than those of any human equilibrist."

Chanute traces the antiquity of rigid-wing aircraft back to the legendary Saracen of Constantinople, who in 1178 attempted to fly from the Hippodrome tower before Emperor Manuel Comnenus. "He stood upright, clothed in a white robe, very long and very wide, stiffened by willow wands," wrote a contemporary. "The Saracen kept extending his arms to catch the wind. At last he rose into the air like a bird."

That he fell and was seriously injured didn't matter to Chanute—he saw in the Saracen's white robes the first design of a rigid-wing MPA.

Such concepts, of course, were forerunners of today's sailplanes, whose motive power comes from gravity and rising air currents. But before soaring craft could become practical, much study and experimentation would have to be done in the areas of wing design to achieve efficient lift/drag ratios, stability about the longitudinal, vertical, and lateral axes, and other aerodynamic refinements.

In the next chapter you will meet some of the pioneers who stripped away earlier concepts of "secret" forces of bird flight. To one delegate to the International Conference on Aerial Navigation held at Chicago in 1893, however, the secret of flight lay in goose feathers.

"How can a 20 pound wild goose carry itself so easily?" he demanded. "Weigh every feather you can pick off from a wild goose, and they will not weigh one pound. Now if the feathers be picked off from a wild goose, he can come no nearer to flying than we can."

It was very simple to him—"a magnetic negative earth-force does the lifting, and that is all produced by the feathers!"

Chapter 4

Why Airplanes Fly

Before man could design and fly a successful man-powered aircraft, a clear understanding of the basic principles of aerodynamics had to be formulated, and some old ideas abandoned. The ancients believed that the atmosphere itself provided a sustaining or impelling force, one that assisted rather than resisted the flight of an object through it.

Leonardo da Vinci did, in fact, recognize that air offered resistance to the movement of a solid object, which he attributed to compressibility effects.

The rapid flapping of a bird's wing, he believed, created a region of compressed air beneath it and so provided lift. He was only partly right.

Similarly, he believed that air flow closing behind a projectile helped to impel it, rather than producing resistance to its flight. (Curiously, Don Luscombe designed the fast little Monocoupe sport plane with a sharply downswept fuselage top surface with this idea in mind—he meant it to act like a wet grape when you squeezed it between your fingers, to make it squirt ahead. Amazingly, it seemed to work.)

Thus misconceptions about aerodynamics have persisted over the centuries. Early investigators, with no wind tunnels at hand, quite naturally had difficulty envisioning what was happening to invisible air at work.

Early French observers were puzzled by the fact that windmills wore out their front axle bearings faster than their rear bearings. There seemed to be some mysterious force in the wind that, under certain circumstances, made an object move toward it rather than away from it. They called this mysterious force "Aspiration."

Captain Le Bris, the French mariner who studied the albatross as a flying model, once wrote: "I took the wing of an albatross and exposed it to the breeze; and lo! in spite of me it drew forward into the wind; notwithstanding my resistance it tended to rise. Thus I had discovered the secret of the bird! I comprehended the whole mystery of flight!"

Professor John Wise, the aeronaut-historian, wrote in 1873 in *Through The Air*: "The philosophers, from Bacon's time down to the discovery of the true nature of atmospheric pressure, as illustrated by the Torricellian tube and air pump, in their speculations upon aerial navigation, all had an opinion that the atmosphere had a defined limit or border, not very high above he earth, upon which the aerial vessel must necessarily be placed in order to have it bouyed up by the air underneath, like the water under a ship. Reasoning from their knowledge of hydrostatics, they took it for granted that the atmosphere was a vast ocean of air surrounding our globe, upon the outer border of which rested another ethereal ocean of a much rarer kind, separate and distinct, as the air rests upon the water."

Of all the early experimenters who were inspired by the example of bird flight to imitate them with mechanical devices, Samuel Pierpont Langley, director of the Smithsonian Institution in Washington, D. C., was first to grasp the notion that MPA designers and builders were going at the problem all wrong.

"Nature had solved the problem of heavier-than-air flight, and why not man?" he asked himself. "Perhaps it was because he had begun at the wrong end, and attempted to construct machines to fly before knowing the principles on which flight rested. I turned to my books and got no help. Sir Isaac Newton had indicated a rule for finding the resistance to advance through the air, which seemed, if correct, to call for enormous mechanical power."

Langley forthwith discarded all the old rules "to commence new experiments, not to build a flying machine at once, but to find the principles upon which one should be built—to find with certainty by direct trial how much horsepower was needed to sustain a surface of given weight by means of its motion through the air."

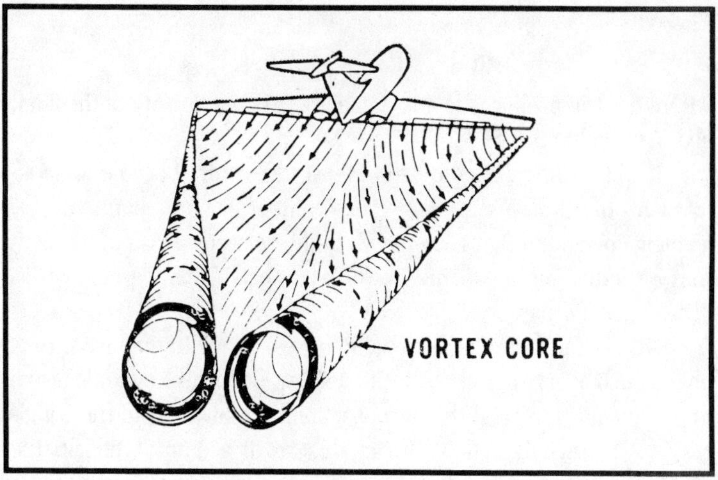

Vortex theory of lift revolutionized aerodynamics.

Long before Langley, other scientists had worried over the problem of air resistance, as opposed to the concept of air "assistance" or "aspiration." Galileo experimentally laid to rest the idea of aspiration and concluded that air resistance was proportional to its relative velocity. Near the end of the 17th Century, the Dutch physicist Christian Huygens went a step further—he showed that resistance of air to the motion of a body was proportional to the square of the velocity.

Sir Isaac Newton's laws of mechanics later formed the basis for classical theories of aerodynamics, based on studies of pressures acting on a flat plate inclined toward an air stream. The pressure, he concluded, was proportional to the product of air density, plate area, the square of the velocity, and the square of the sine of the angle of inclination. Leonhard Euler refined the formula, stating that the effect of the plate's inclination was proportional to the sine, not to its square. But, none of these early experiments revealed the effects of flow over the upper surface of the plate, where low pressures helped to produce the major portion of a wing's lift.

Not until much later would the bigger picture emerge, of total circulation flow, and of air as a continuum, with a pressure field extending for considerable distances from the plate.

By bits and pieces, the jigsaw puzzle picture of how air works, and how an airplane flies, became clearer. In the mid-18th Century, Daniel Bernoulli, a Swiss physicist, arrived at his well-known theorem:

$$q + \frac{1}{2} PV^2 = \text{constant}$$

where q is the pressure in an airstream, P is the density of the fluid, and V its velocity.

Jean Le Rond d'Alembert in the same century became dissatisfied with the theories for resistance, and after some mathematical juggling concluded that resistance, being related to loss of momentum on impacting a moving body, was zero. This was known as "D'Alembert's paradox."

Others tackled the relationship of resistance to the viscosity of a fluid such as air or water. In the 1880s, Osborne Reynolds discovered laminar flow and the abrupt transition, away from the wing's surface, to turbulent flow. From this work a new yardstick, the "Reynolds Number,"evolved. It expressed the ratio of mass forces to viscous forces in a fluid, and enabled designers to compare aircraft models to full scale machines in airflows of equal Reynolds Numbers.

Researchers now had a better understanding of air at work, and mathematical formulae with which to grasp the meaning of measurements inside wind tunnels. But not until the beginning of the 20th Century did a fuller understanding of total circulation appear.

In England, Frederick W. Lanchester set forth the circulation theory of lift of an airfoil of infinite span, and the concept of a vortex theory of the lift of a wing of finite span. Coincidentally, two other researchers came to the same understanding and provided a mathematical theory to explain it. They were M. Wilhelm Kutta in Germany, and Nikolai E. Zhukovski in Russia.

Another brilliant researcher, Ludwig Prandtl, at Gottingen, Germany, independently arrived at the same concept, forming the basis of modern aerodynamic theory. Prandtl, incidentally, was first to recognize that total resistance of an aircraft consists of three parts—skin friction (viscosity), form or pressure drag, and induced drag.

Understanding the latter form of drag, resulting from production of lift by airflow over a wing, revolutionized aerodynamic theory. It was apparent that induced drag is smaller as the wingspan increases, primarily because it is associated with the vortices trailing from each wingtip.

Here was a fresh starting point for designers of manpowerea aircraft—because the human machine is low powered, drag had to be

Octave Chanute's tri-decker hang glider of 1896.

Chanute experimented with as many as four and five multiplane gliders.

minimized. And as induced drag is the bigger part of total drag, a longer wingspan is essential.

A lack of theoretical understanding of aerodynamics in the 19th Century did not deter MPA experimenters from pushing ahead with aircraft patterned after wings of birds. More serious researchers like Professor Langley tested their designs in model form, whirled through the air at the end of rotating beams.

Both in the United States and abroad, efforts were made to get together all the existing knowledge of mechanical flight for serious study. Thus in January, 1866, was born the Aeronautical Society of Great Britain, and two years later it sponsored the first Aeronautical Exhibition, at the Crystal Palace at Sydenham. A total of 77 exhibits were on display for eight days, and MPAs were predominant.

In America, Dr. Albert F. Zahm, a Notre Dame University physics professor, invited the world's leading aeronautical experimenters to gather at the Chicago World's Fair grounds in 1893 to share ideas. Octave Chanute at first refused to chair the Conference

Chanute multiplane glider in flight.

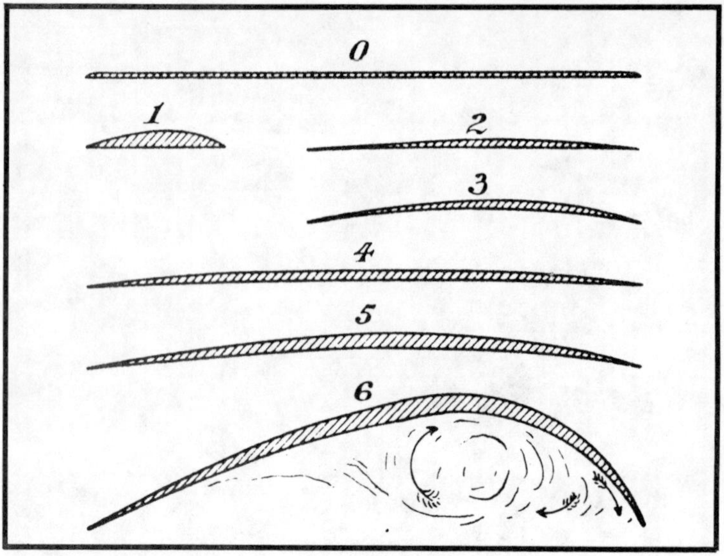

Horatio Phillips patented a series of airfoils in 1891. Number 5 seemed to work best.

on Aerial Navigation, fearing an invasion of "cranks." He finally accepted the honor—his publication of *Progress in Flying Machines* in 1891 made him the logical choice.

Inspired by the Crystal Palace exhibition, Horatio Frederick Phillips launched a study of wing curves in a crude wind tunnel, demonstrating that flat plates were useless for supporting weight in air. In 1884 Phillips patented a series of airfoils of concave-convex shape. In his patent description, Phillips noted that "particles of air

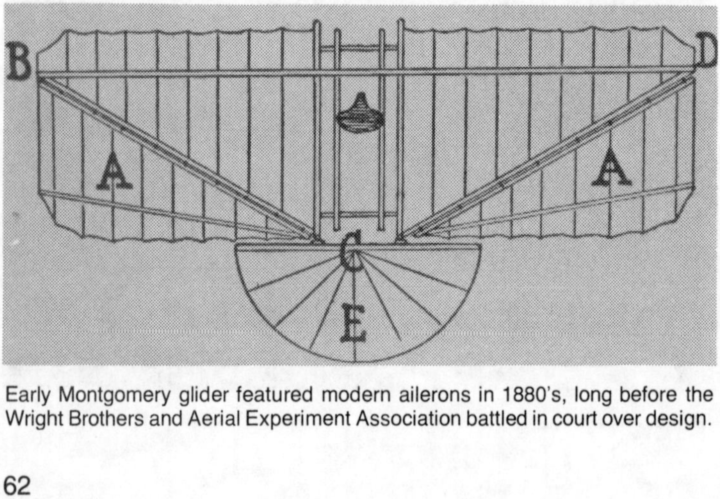

Early Montgomery glider featured modern ailerons in 1880's, long before the Wright Brothers and Aerial Experiment Association battled in court over design.

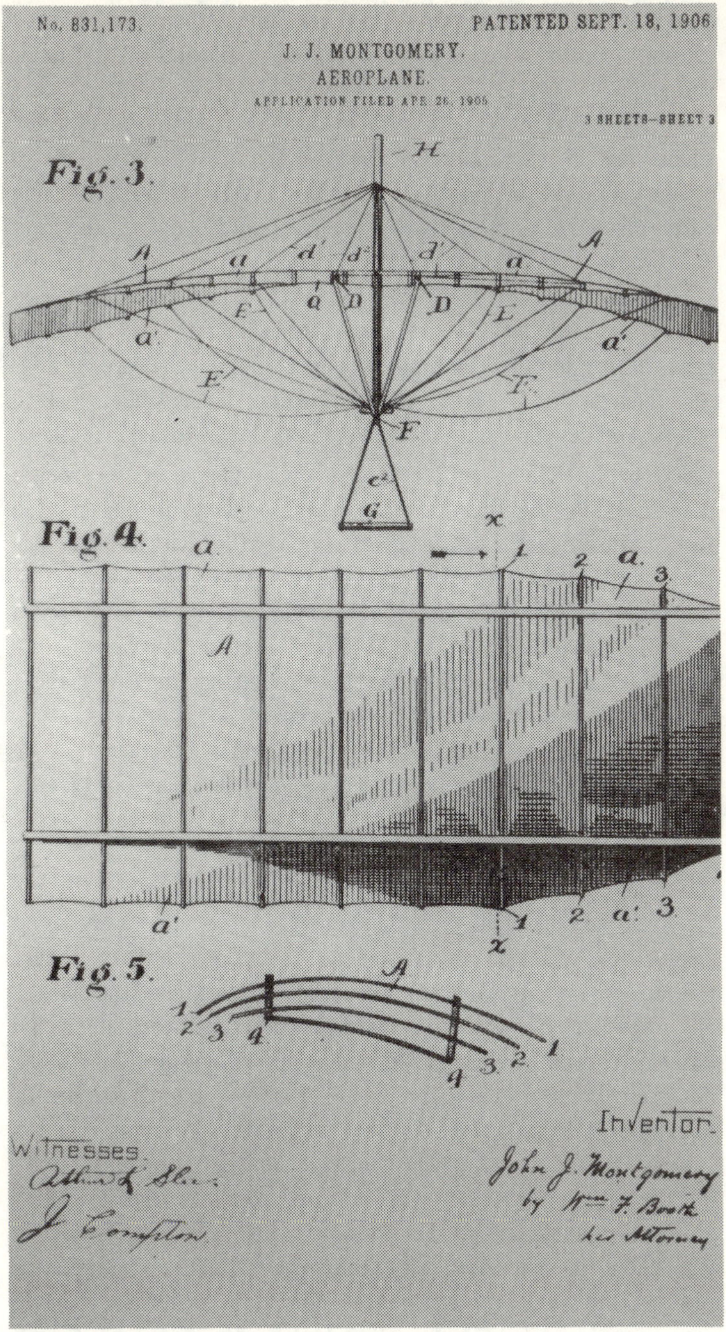

John J. Montgomery won patent on curved wing aeroplane in 1906.

struck by the convex upper surface are deflected upward, thereby causing a partial vacuum over the greater portion of the upper surface."

Professor Langley drew criticism at the Chicago meeting by suggesting, in his paper, *The Internal Work of the Wind*, that a mysterious force within the wind in some way pulsated, allowing birds to soar. Chanute, however, agreed with Langley—to him it was simply "aspiration."

There were startling developments at Chicago—at least two papers proved that successful heavier than air flight already had been achieved, by John J. Montgomery and Louis P. Mouillard.

Back in 1884, Montgomery revealed, he had been making 600-foot glides down a sandy slope at Otay Mesa in Southern California. In 1883 he had failed to get a wing-flapper off the ground, then designed a rigid-wing craft with pivoted wings that again failed to fly. Subsequently he designed wings, fitted with rudimentary ailerons, but they were flat wings.

After another six years of independent study, he developed a parabolic curve that worked well. To Montgomery, it seemed the curved surface deflected the air downward, so that the wing in effect rode the crest of a wave, like a surfboard.

What amazed Chanute was how he developed the curve. With no knowledge of how to construct a wind tunnel, he borrowed a feather pillow from his mother's linen closet, set the wing up on a hilltop in a high wind, and watched the way the feathers blew over and around it.

Next he built small tin wing sections and mounted them inside an open box, with a damper at one end. He set up a mirror to reflect the sun's rays at right angles to the airflow, then opened the damper and sifted in smoke and dust particles. The crude heliostat enabled him to see both laminar and turbulent airflow patterns, and so he designed a curve that gave the best L/D ratio—the most efficient lift for the least drag.

In Cairo, Egypt, Mouillard worked along different lines, studying the flight of great African vultures to determine how they achieved controlled flight. He designed a pair of ailerons for a rigid wing glider that worked the way the vulture's feathers worked. To turn left, the left aileron was depressed into the airstream, creating drag that effected a turn. In 1892 Mouillard applied for a United

Prof. John J. Montgomery (third from left) poses with his tandem hang glider Santa Clara. Pilot Dan Maloney (far right) was killed attempting aerobatic maneuver after a balloon launch.

Replica of Montgomery's tandem glider flown for movie "Gallant Journey" proved design was successful.

States Patent for his aileron control system. It was duly granted May 18, 1897.

Interestingly, the secret of the ultimate success of Dr. Paul MacCready's Gossamer Condor lay in just this principle, of using wingtip drag to assist in a turn, at the same time providing extra lift by increasing the relative angle of attack of the inner wing.

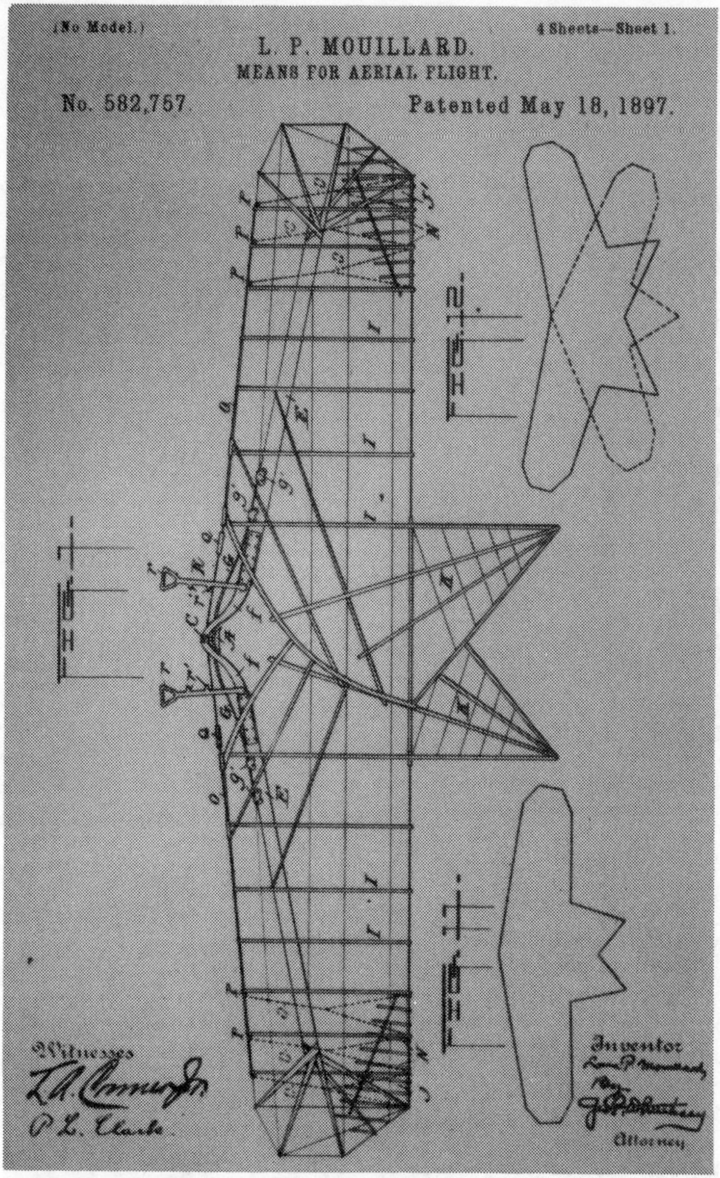

In 1897 French designer Louis P. Mouillard patented glider design featuring wingtip "dragerons" used today in modern hang gliders and Gossamer Condor MPA.

As stability became more important in aircraft design, legal problems arose. Orville and Wilbur Wright fought bitterly through the courts to establish priority for their claim to have invented the

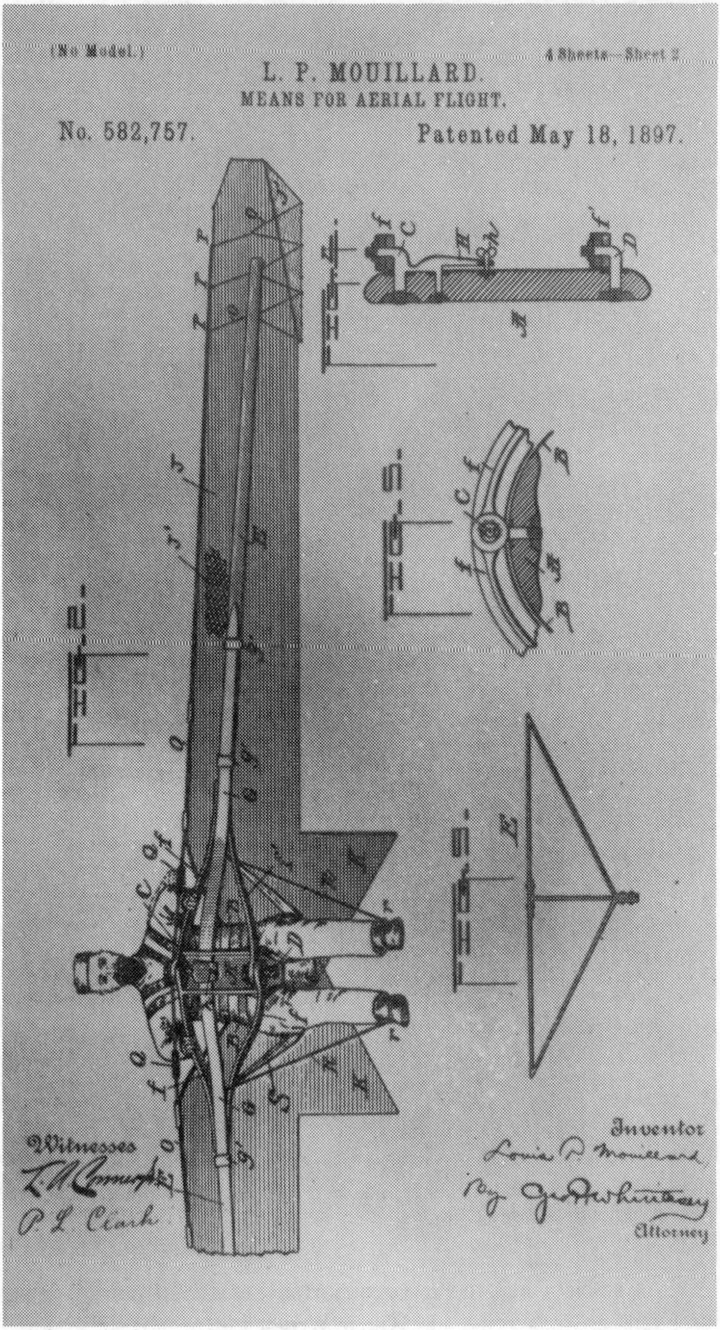

Patent drawing of Mouillard's "Drageron" control system.

"three-torque" system of control of an aircraft, using a combination of wing-warping, elevator, and rudder surfaces.

Injunctions were filed to prevent other birdmen like Louis Paulhan from flying in this country, but the chief "enemy" of the Wrights was the Aerial Experiment Association. Alexander Graham Bell, an AEA member, had invented and patented a three-torque control system that employed hinged ailerons, not wing warping. It was flown successfully by Glenn Curtiss and other AEA members.

But still earlier, according to Dr. Zahm, was a three-torque control system invented by the French experimenter Alexander Goupil in 1884, and flight tested in January, 1917, by the Curtiss Aeroplane and Motor Corporation, at the peak of the patent fight.

Even the Goupil system was antedated, Dr. Zahm pointed out, by a British patent issued in 1868 to Matthew Boulton.

Today, such developments in the history of mechanical flight serve only to remind us that man, over the centuries, has created a technical heritage that has taken us to the moon, and into hypersonic atmospheric flight. In the realm of low-velocity man powered flight, however, the search for new ideas goes on.

Chapter 5
Onward and Upward

Otto Lilienthal, killed in a glider crash in 1896, might have lived well beyond his 48 years had he followed his own advice and stuck with man powered aircraft, instead of trying to fly an ornithopter with movable wingtips and a carbonic acid gas motor. He had made more than 2,000 successful glides before the urge to switch to power sidetracked his amazing path of progress.

Some 30 years of research and preparation by Otto and his brother, Gustave, preceeded his first glider flight in 1891. Lilienthal's early gliders strongly resembled the onithopter batwing designs of Leonardo da Vinci, his first one weighing 40 pounds, with a wing area of 107 square feet, made from waxed cotton cloth stretched over a framework of peeled willow wands.

From the outset he was aware that controllability was the big problem in man powered aircraft. He once cried: "To fly is everything! To contrive a Flying Machine is nothing; to construct one is something; to control it in flight is to reach the heights."

For starters, Lilienthal practiced flight control with a machine of Multiple sets of wings, operated by foot power, and suspended from a boom projecting from the roof of a house. His wings were gracefully curved, and when flapped raised him 30 feet off the ground, assisted by counterweights at the end of a rope run over a pulley.

Otto Lilienthal, famed German glider pilot, died in this crash on August 9, 1896.

In 1889 Lilienthal came to a number of conclusions, published in his *Der Vogelflug als Grundlage der Fliegekunst* (Bird Flight as the Basis of the Flying Art):

- Practical flying machines in nowise depend upon the discovery of light and powerful motors.
- An ordinary man's strength is sufficient to work efficiently an appropriate flying apparatus.
- The wings of large, high-flying birds should be used as models.
- Experiments must determine whether to use the broad wings with spread-out primary feathers of waders, or narrow, tapered wings of sea birds.
- Wings must be curved, concave on the bottom side.
- Depth of the curve should be one twelfth its chord.
- Wing spars should be located as far forward as possible.
- In flapping flight, the widest portion of the wing must cooperate in the upstroke to sustain weight.
- The wingtips only need oscillate, the rest remaining rigid for soaring flight.

Lilienthal had workmen build a gliding hill, from dirt removed for a canal near his home at Gross Lichterfelde, near Berlin, with a launch platform constucted on top and a cave dug below to store his gliders.

To test his ornithopter wings, he also installed a springboard in his garden. After a sprint of 26 feet, he bounced into the air, learning how to flap. "Dexterity alone," he once wrote, "invests the native inhabitant of the air with superiority over man in that element."

On August 9, 1896, Lilienthal made one final glide from his artificial hill, to test a new rudder control, a line that ran to a head band. It was to be used next day in a powered glider, that had proved to be unstable. On that flight Lilienthal's machine stalled and crashed. He died the next day of a broken back.

If Lilienthal had given his life in an effort to wrest the secret of muscle-powered flight from the birds, it was not in vain. He would become the inspiration for future experimenters, including the Wright Brothers, and in a sense proved that earlier ornithopter advocates like da Vinci had not been altogether wrong.

Over the centuries since da Vinci dreamed of flying from Swan Mountain, the subject of human flight became an important part of

Lilienthal, like da Vinci, used batwing design, with a cruxiform tail.

Cabane struts and wires supported batwings of Lilienthal glider.

Lilienthal flies biplane glider to base of hill.

English literature. A common theme was based on efforts of a shipwrecked mariner to escape from an island, a plot immortalized by Daniel Defoe's *The Life and Strange Surprizing Adventures of Robinson Crusoe*, which appeared in 1719.

Robinson Crusoe, based upon the real-life survival story of the shipwrecked Scottish sailor Alexander Selkirk, was preceded by 81 years with the publication in 1638 of Francis Godwin's *The Man in the Moone*, another island adventure tale. In this story the hero, Domingo Gonsales, a Spaniard, is abandoned (by reason of sickness) during a sea voyage, on the then uninhabited island of St. Helena.

The escape theme is not so mundane as Crusoe's efforts at building a raft. Instead, Gonsales occupies his time training a number of wild swans to obey his call, and eventually to fly with small burdens.

Eventually he trains a flock of seven "Gansas or Large Geese" to fly in harness carrying a small lamb, which he envies as "the first living creature to take possession of such a device." (Prophetically, this reference was followed on September 19, 1783, 65 years later, by the ascension of a sheep, a cock, and a duck, in a hot-air Montgolfier balloon basket, as history's first living creatures to ascend in a manmade aerial vehicle.)

The story gets a bit wilder as Gonsales hooks up 25 Gansas and goes aloft himself. Later, when he is shipwrecked returning to Spain, Gonsales is rescued by his Gansas, who head for the moon—their real home—and carry him thence in 11 days, on a migratory flight. On the journey, the birds encounter gravitational equilibrium and remain stationary for a while, resting, before continuing their space flight.

The next author to combine the shipwreck theme with human flight escape was Robert Paltock, in 1751, in a book whose title page tells the whole plot: *The Life and Adventures of Peter Wilkins, a Cornish Man: Relating particularly, His Shipwreck, near the South Pole, his wonderful Passage thro' a subterraneous Cavern, into a kind of New World: his there meeting with a Gawry or flying woman, whose Life he preserv'd, and afterwards married her; his extraordinary Conveyance to the Country of Glums and Gawrys, or Men and Women that fly... Taken from his own Mouth, in his Passage to England from off Cape Horn in Africa, in the Ship Hector.*

There is more to the story, of course—Peter Wilkins, defending the kingdom of Doorpt Swangeanti (the Land of Flight) against a

Lilienthal launching from platform atop his flying hill.

rebel invasion by 5,000 winged men flying in five layers, opens fire and shoots down 300. "They fell so thick about me," he relates, "that I had enough to do to escape being crushed to death by them."

In the same year that *Peter Wilkins* appeared, the Rev. Ralph Morris brought out another shipwreck flying story entitled "The narrative of the Life and astonishing adventures of John Daniel, a smith, at Royston in Hertfordshire."

Here, the hero's son, also mechanically minded, builds an ornithopter with several ribbed wings "cloathed with callicoe dipt in wax" and operated by pump handles. He took his father for a ride and "away we went," the narrator says.

We now come to Dr. Samuel Johnson, the noted British writer and lexicographer, who reportedly became interested in man-powered flight when he once roomed in the same lodging house with a man who broke his legs in a fall while attempting to fly. In a ficticious letter to a British periodical, The Rambler, in 1752, Dr. Johnson, signing himself "Hermeticus," reports that he had twice dislocated his limbs and once fractured his skull trying to fly.

Seven years later he devoted a full chapter to the subject of man-powered flight in a hastily written book, *Rasselas*, which he turned out in a single week and sold for 100 pounds to defray the costs of his mother's funeral. Rasselas is an Abyssinian prince of eloquent speech, who discourses on flight with a young artist building a flapping-wing flying machine.

The artist seems to have had a good grasp of the mechanical, if not the aerodynamic, requirements of heavier-than-air flight. He says: "You will be, necessarily, upborne by the air, if you can renew any impulse upon it, faster than the air can recede from the pressure." (In 1861, J. S. Phillips used these words almost verbatim in an abridgement of the patent specification for his invention of a Flying Machine).

When Rasselas complained that "every animal has his element assigned him; the birds have the air, and man and beasts the earth, the artist had a ready answer: "So, fishes have the water, yet beasts can swim by nature, and men by art. He that can swim need not despair to fly; to swim is to fly in a grosser fluid, and to fly is to swim in a subtler. We are only to proportion our power of resistance to the different density of matter through which we are to pass."

If the artist seems to have had a good grasp of fluid mechanics, his ideas on gravity and how to escape it were far-fetched: "The

Lilienthal perfected use of "Body English" for lateral balance.

labor of rising from the ground," he says, "will be great; but as we mount higher, the earth's attraction and the body's gravity will be gradually diminished, till we shall arrive at a region where a man will float in the air without any tendency to fall: no care will then be necessary, but to move forwards with the gentlest impulse.

"You, Sir," he tells Rasselas, "whose curiosity is so extensive, will easily conceive with what pleasure a philosopher furnished with wings, and hovering in the sky, would see the earth and all its inhabitants rolling beneath him, and presenting to him successively, by its diurnal motion, all the countries within the same parallel.

"How must it amuse the pendent spectator to see the moving scene of land, ocean, cities, and deserts! to survey, with equal security, the marts of trade, and the fields of battle; mountains infested by barbarians, and fruitful regions gladdened by plenty and lulled by peace! pass over distant regions, and examine the face of nature from one extremity to the other!"

When Rasselas complained that the upper atmosphere might be too thin to breathe, the artist volunteered to make the first test flight of the wings by himself. Rasselas agreed, and for a year dropped by from time to time to check on his progress.

Finally the wings were finished, Johnson wrote, "and on a morning appointed, the maker appeared furnished for flight on a promontory; he waved his pinions awhile to gather the air, then leaped from his stand, and in an instant dropped into the lake. His wings, which were of no use to him in the air, sustained him in the water, and the prince drew him to land half dead with terror and vexation."

If Dr. Johnson's vision of man-powered flight was futuristic, he did live to hear of man's conquest of the sky in 1783 by hot-air balloons, before his death in 1784. The word came in letters written to Sir Joseph Banks in London, from Benjamin Franklin in Paris. Franklin, the American Ambassador to France, was a serious supporter of aerostation. No doubt Franklin had read Francis Godwin's *The Man in the Moone* and Robert Paltock's *Life and Adventures of Peter Wilkins*, both of which employed harnessed birds to provide lift and propulsion, for among Franklin's papers, in the American Philisophical Library, is a curious drawing of a conical balloon to which 16 birds are harnessed.

Among the more amusing early references to the ornithopter principle of flight was an article which appeared in the Scientific

Flock of wild geese hooked to flying machine provided power for fictional flight to the moon, 17th Century.

American in 1851 on "A successful experiment with a new apparatus for flying."

"The flyer," according to the article, was a Miss Juanita Parez, who though rather fat and corpulent, moved through the air, by the help of wings, with the greatest of ease and rapidity. She was

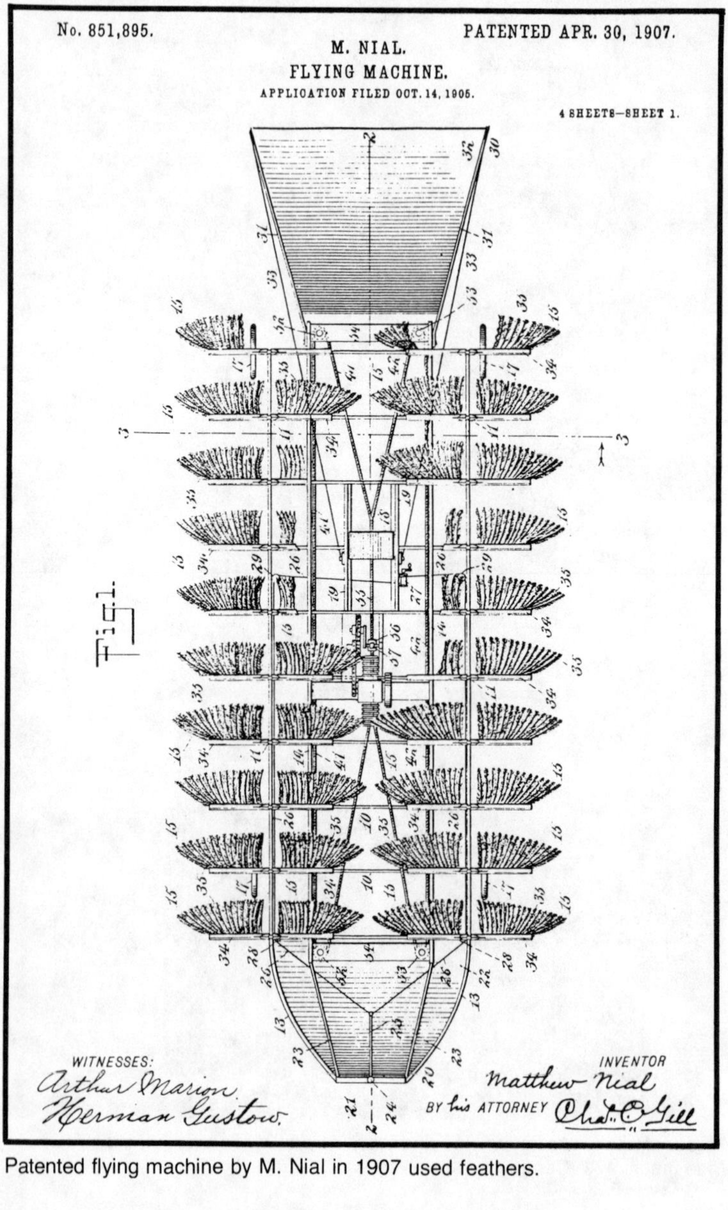

Patented flying machine by M. Nial in 1907 used feathers.

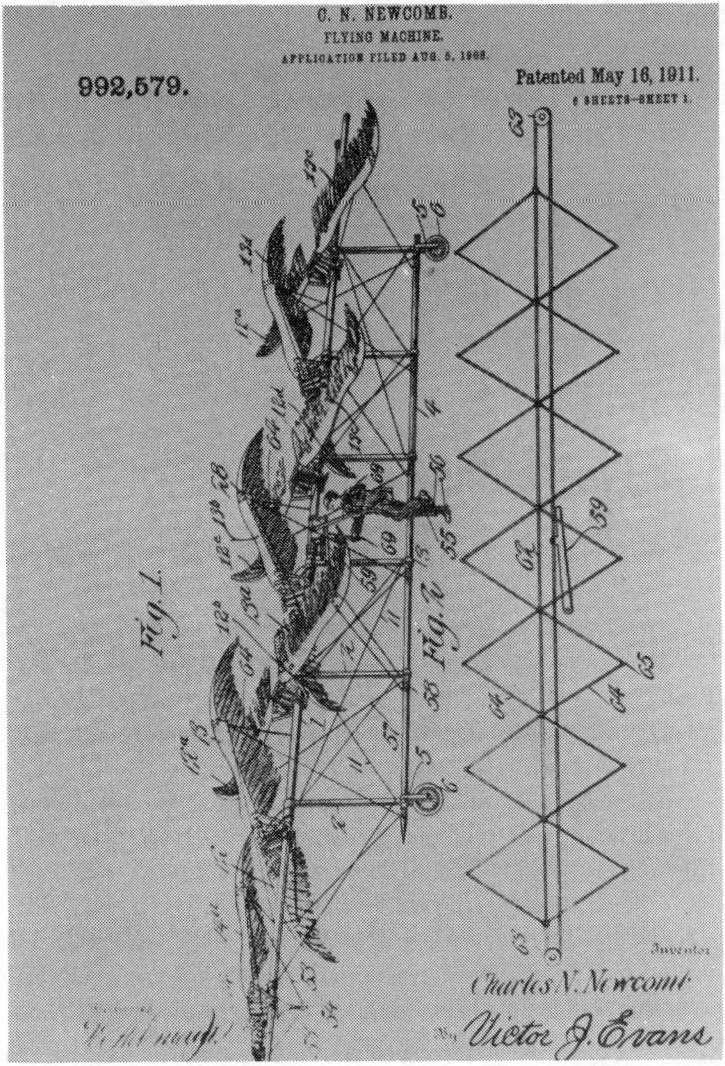

C. N. Newcomb's Flying Machine used man power to flap feathered wings.

advertised to fly a distance above 1200 feet, raising above 600 feet, but exceeded the program both in height and distance. No description of the wings is given. They have a spread of some fifteen feet, are fastened by ligaments of great flexibility, and arranged so as to move with great rapidity, and they make a noise like a windmill. The astonishment at Madrid was immense, and no wonder—just think of a corpulent damsel flying through the air and making a noise like a windmill!"

Chapter 6

The Westerners

"He regretted that he was not a bird, and could not be in two places at once."

—Sir Boyle Roche (1743-1807)

Reuben Jasper Spalding, a Colorado prospector, spent seventeen years of his hard life working on a lovely spread of feathered wings, which he dreamed of soaring amongst the great eagles near his cabin at Rosita.

Like others before him, Spalding saw no reason to believe that he could not fly with them, rising on the slope winds that carried the wheeling birds up the face of the eastern escarpment, sometimes even above the timberline and into snow country.

Built in secret, first word of the Spalding Flying Machine leaked out when William J. Orange, enterprising editor of the *Silver Cliff Russler*, sneaked into the miner's barn one day in 1890 and wrote up a feature story about what he saw.

Forty-six years old when he began work on his winged wonder, Spalding was 63 by the time it was ready to fly. Rather than being angered by Orange's story, he felt a measure of pride that he had been noticed.

"The machine is fashioned exactly like a bird's wings and tail," Orange had written. "The wings and tail and all the springs, hinges, cords and attachments to operate them are attached to a leather harness. The steel used is of the best quality, having cost 80 cents a pound."

In each wing, Orange counted, were 14 feathers made of silk and eagle quills. There was a head rest and leg rests and the "flapping, curving and folding of the wings is controlled by an easy forward arm motion."

The workmanship was, in fact, so gorgeous that the patent commissioner, in issuing Patent No. 398,984 to Spalding on March 5, 1889, called it "the only perfect duplicate of nature's flying apparatus ever made in the world," Orange wrote.

But would it fly?

H. F. Moore, a resident of nearby Florence, Colorado, told the author some years ago: "Spalding would get up on top of buildings and fly to the ground."

In the summer of 1888 the old miner went to Denver and took a train clear to Washington to arrange for his patent. He carried a model tucked under his arm, arousing much curiosity and considerable laughter. In the capital, he sat for the only portrait known to exist of him.

Back home, Spalding arranged to give a flight demonstration of his machine at Denver. Shops were closed. Cowhands and miners came to the city from miles around. Old Reuben strapped on his lovely wings and jogged up and down the street to limber up.

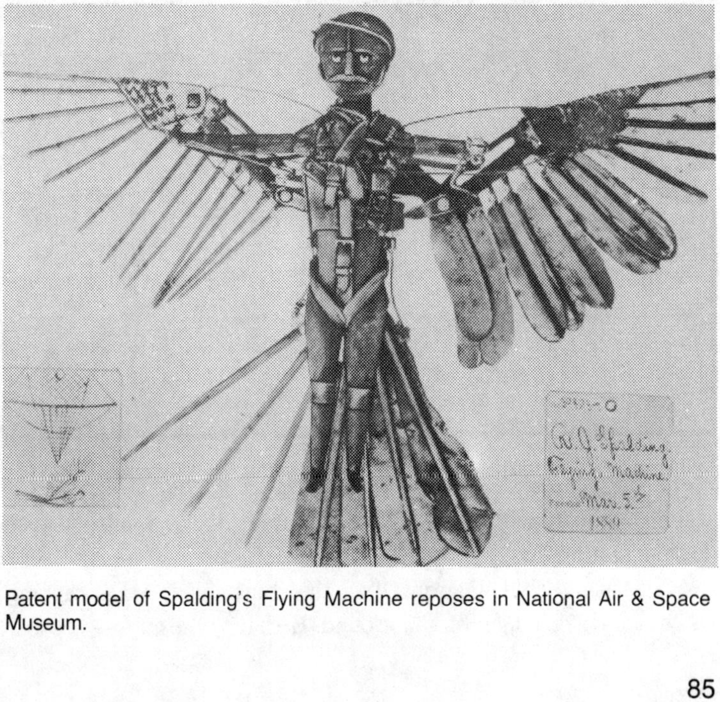

Patent model of Spalding's Flying Machine reposes in National Air & Space Museum.

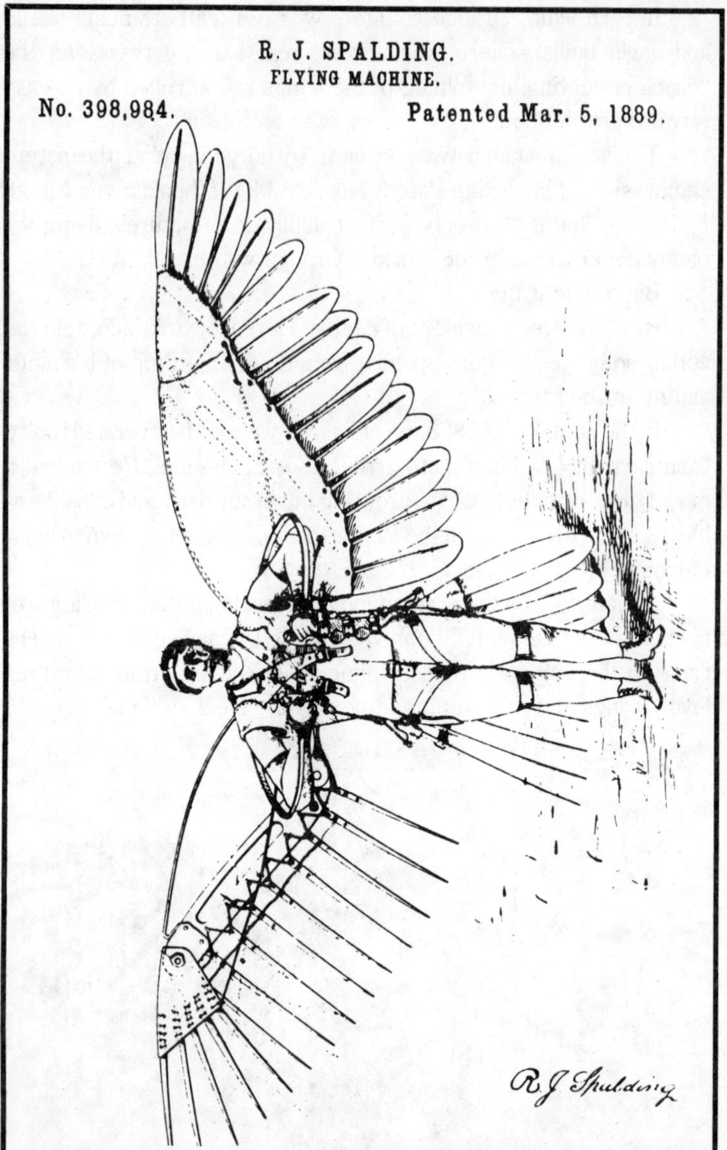

Front view of Spalding Flying Machine shows springs used to assist in flapping.

"Go ahead and fly!" someone shouted. A gun went off; dust exploded at his feet.

In alarm, Spalding sprinted for the courthouse, wings flapping as he ran. He didn't get off the ground. He hadn't meant to—not that way. He paused for breath, waiting for friends to show up with a hot-air circus balloon to lift him into the sky, so that he could cut loose and glide back to earth.

The crowd of cowhands ran after the old man, grabbed him from behind and tossed him roughly into the air, amid shouts of laughter.

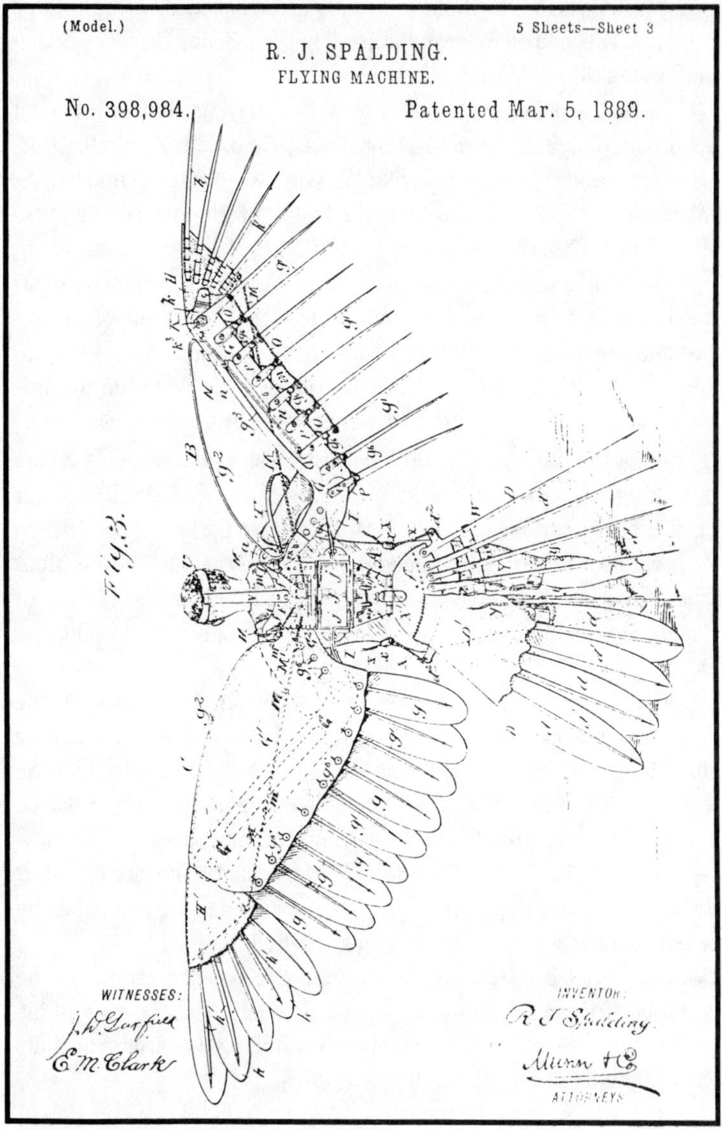

Rear view of Spalding Flying Machine shows coiled springs (j) and method of attaching feathers.

But Spalding was tough. Eyes flashing, he lit into the crowd like an avenging angel, kicking and punching.

When the street brawl ended, Spalding's lovely wings hung loosely, the parts broken. The leather body harness bore marks of sharp Bowie knives.

"I afterwards saw those knife cuts in the harness, after they had been neatly repaired by grandfather," his grandson, Clarence Spalding, once told the author.

Sadly the old man packed away his angel wings and forgot about his dream of flying. He died at Swallows, Colorado, April 19, 1902, the same year Wilbur and Orville Wright, two Dayton, Ohio bicycle mechanics, were preparing to make history with powered flights at Kitty Hawk, North Carolina.

Spalding's wings were destroyed in a fire in 1917, but two of his hand-crafted patent models remain to show what an old man's dreams are made of. One reposes in the National Air and Space Museum in Washington, an official tribute to a forgotten aviation pioneer. The other was donated to the author by Clarence Spalding.

A contemporary flying enthusiast who shared Spalding's dream of soaring high above the Old West was Sure Shot Bill Beeson, a familiar character around Dillon, Montana faro parlors in the 1880s.

A carpenter by trade, Beeson had joined the rush to the Montana mining country along the Beaverhead River, in the Bitterroot Range. He decided to pursue his trade building houses in booming Dillon.

For recreation, old timers recall, Beeson played faro by the hour, stroking his moustache and watching the cards with calculating shrewdness. Because the house seldom got the better of him, he won the sobriquet "Sure Shot Bill." George R. Metlen, a Dillon mining engineer and former highway commissioner, who knew Beeson well, recalled that "he would put a dollar on the faro layout. If he lost the dollar he quit, but if he won he stayed in the game until he began to lose, when he immediately cashed in."

Sure Shot Bill Beeson had another vice besides gambling—he was a born Darius Green. He kept a pigeon loft at the rear of his workshop, and would spend long hours watching the birds gracefully wheeling and dipping through the sky.

In 1881, with assistance from the Scientific American Patent Agency, which did a thriving business with independent inventors,

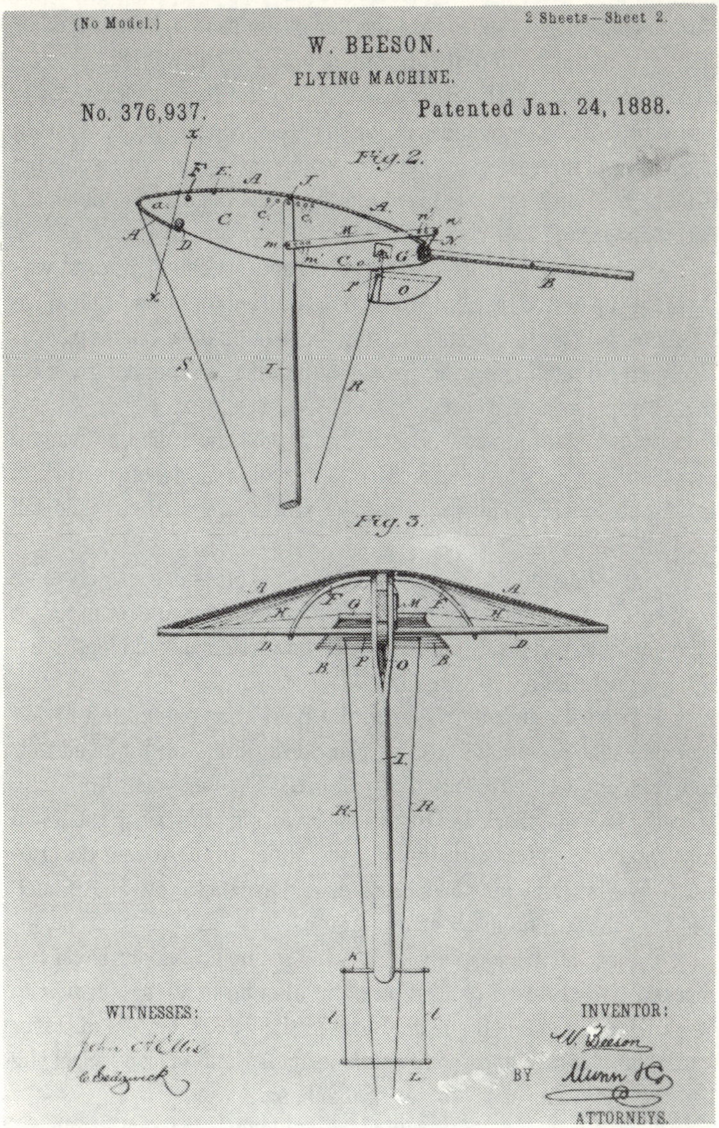

In 1888 Beeson patented Flying Machine with tail operated by pilot riding on a trapeze.

Sure Shot Bill patented a Flying Ship that was just what its name implied. Utterly simple in concept, it consisted of a boat hull from which two masts protruded upward and outward. Sails hung from crossbars between the masts, like sheets on a clothesline. A system of ropes and pulleys permitted the set of the sails to be changed in flight to catch the wind at the proper angle. That was all.

"To start the ship," Beeson explained to the Patent Office, "it is carried up by a balloon, and is dropped at the proper time, or is raised like a kite, all the sails being set almost vertically—and when the ship is up high enough this rope is cut in the same manner as one would cut a kite string."

When he wasn't playing faro or pigeon-watching, Beeson was dreaming up new ideas. In 1887 he patented another new brainstorm for Aerial Navigation. It also consisted of a number of canvas planes, arranged to catch the wind from any quarter, suspended by rope from a supporting balloon.

Below his "four-winged sail" hung what he called a "winged propeller"—an enclosed canvas cell in which an aeronaut rocked from side to side, causing a pair of fabric wings to rise and fall alternately.

And below the winged propeller was his piece de resistance—a "sail suit"—two sheets of canvas sewn together with room for an aeronaut to crawl inside, where he could work arms and legs like a flying squirrel in an effort to fly.

Not one to give up, Beeson the next year patented a Flying Machine that resembled a huge cambered kite with a hinged tail, from which was suspended a trapeze. The aeronaut rode the trapeze, swinging back and forth and so working the tail up and down for propulsion. As an afterthought, he added in the patent description: "The machine may also be made and used as a toy, and as such will afford amusement to many persons."

Typical of the lone inventor who stumbled blindly along unmarked trails toward realization of an idea he knew simply *had* to work, Beeson plunged ahead. Early in the 1890's he took a closer look at his pigeons, and decided that perhaps Reuben Jasper Spalding had the right approach—a set of mechanical wings.

"His machine was a rather ingeniously constructed affair," George Metlen wrote to the author years ago. "It was shaped as near the form of the pigeon's wings as he was able to build it out of thin strips of pine wood. The wings were about six feet in length by 18 inches in width at their widest part."

Early in 1892 Beeson felt the wings were properly adjusted and braced for an actual flight test. For an assistant he called on his

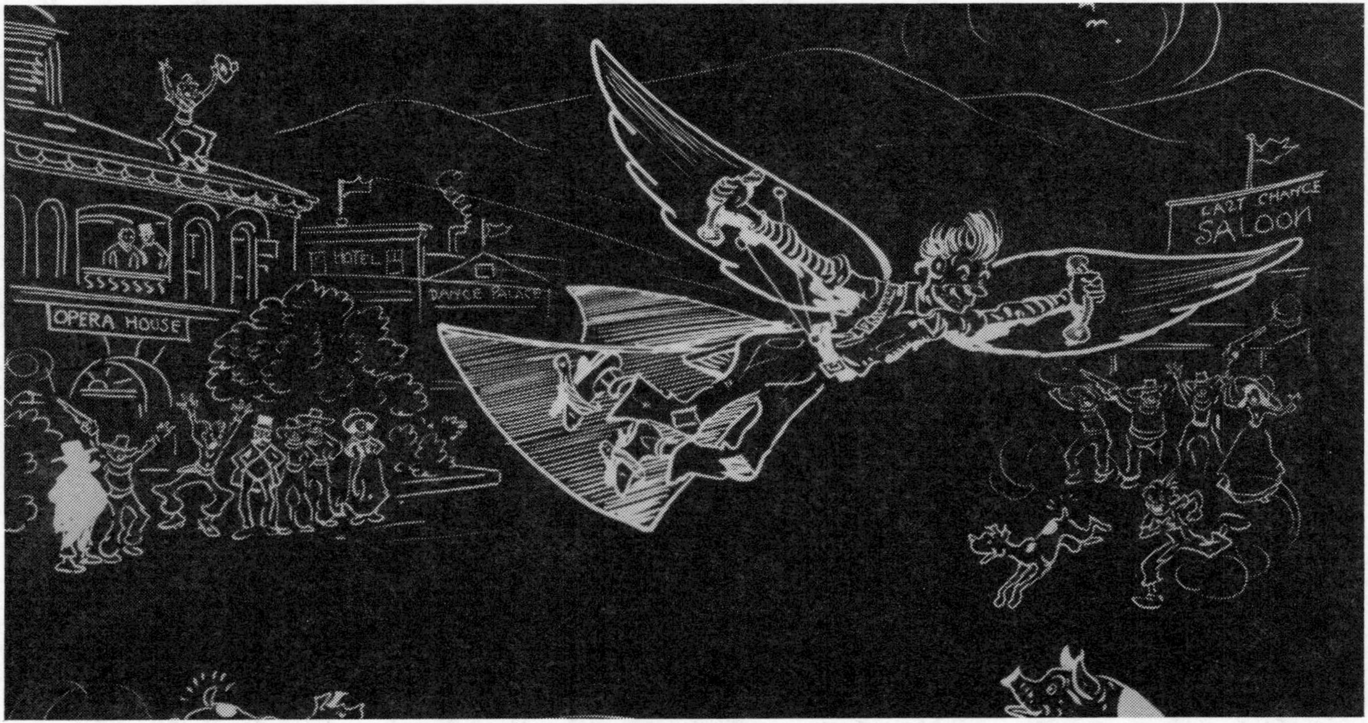

Author's father, cartoonist Clare Victor Dwiggins, did this sketch of Sure Shot Bill Beeson's famous flight from Opera House roof in 1880s.

favorite faro dealer, a person known as Billy the Dick. Together they carried the wings to the roof of Dillon's two-story opera house, some forty-five feet above the street. The whole town turned out to watch.

Billy the Dick strapped Sure Shot Bill into his flying machine, and after a few experimental wingflaps, Beeson tensed his muscles for the leap that would bring him fame or misfortune. Metlen, who was there, describes the great Dillon flight:

"After the straps were all securely fastened, Bill took off into space. The results were not wholly satisfactory; however, he was on the right track.

"The wings acted like a glider and he planed to the ground about two hundred feet from the take-off. The air resistance was so great that he could not flop his wings, and had it not been for the tie-down, he would have fell directly to the ground; but instead, he made a gliding, sliding landing, the buck stay on his belly landing first and with enough force to about wreck old Bill and knock all the air out of him. One wing was reduced to splinters. Notwithstanding the jibes old Bill was subjected to by the crowd that gathered to witness the flight, he did actually glide from a height of 45 feet a distance of 200 feet from the take-off to the ground."

Beeson still would not give up, Metlen recalled, for "After his unsuccessful flight he built another machine along the same lines as the first, but tried to operate it by pedaling the wings from crank pedals like a bicycle. He was able to get up enough speed on his wings to make the contraption jump from the floor of his shop, but she lacked the necessary horsepower to raise off the ground."

Chapter 7

The Tunnel Men

"They have sown the wind, and they shall reap the whirlwind."
—Hosea viii:7.

Introduction of the wind tunnel as an aerodynamic research tool was an important step toward understanding basic laws governing not only the flight of birds, but of all heavier-than-air craft which depend upon the atmosphere for support. As Mouillard put it, observing bird flight was one thing; understanding it was something else.

Whereas the 1893 Chicago Conference on Aerial Navigation convinced the scientific community that they were on the right track to achieving mechanical flight, little was understood of the basic principles of aerodynamics involved. It was well and good to watch birds fly, but there had to be some way to simulate flight in a controlled laboratory environment, before accurate measurements of lift, drag, and other forces could be achieved.

Oddly, many of the foremost researchers involved in aerodynamic studies continued their bird-watching in hope of learning the secret of flight.

Professor Samuel Pierpont Langley observed: "Nature has made her flying machine in the bird, which is nearly a thousand times as heavy as the air its bulk displaces, and only those who have tried to rival it know how inimitable her work is, 'for the way of a bird in the air' remains as wonderful to us as it was to Solomon."

Sir Hiram Maxim said: "Man is essentially a land animal. It is quite possible if Nature had not placed before him numerous examples of birds and insects that are able to fly, he would never have thought of attempting it himself."

Both Langley and Sir Hiram, however, eventually put aside their bird-watching and turned to the laboratory to understand the dynamics of flight. The old empirical laws of mechanics somehow didn't work.

As early as 1808, Sir George Cayley in England had come remarkably close to grasping the mechanical action of a bird's wing, in a sketch of an ornithopter that clearly shows the bird-like action of a wing's leading-edge spar and its flexible area behind, in creating a propelling action on both the upstroke and downstroke.

The British aeronautical historian Charles H. Gibbs-Smith, in his book: *Sir George Cayley's Aeronautics 1796-1855*, (Her Majesty's Stationery Office, London, 1962), described this action with the interesting explanation:

"All birds are propelled by the airscrew action of the outer parts of their wings, which provide all the thrust and some of the lift: in some birds the emarginated outer primary feathers of each wing twist into four or five airscrew blades; in some high-aspect-ratio birds the outer primaries twist collectively into a single blade; and in birds with wings of large area, and medium aspect-ratio, nearly half the wing is twisted in this way. As all the movement of a wing is communicated by the equivalent of a leading-edge spar, the outer hind portion heels up on the down-stroke, and down on the up-stroke; but owing to the camber of the wing construction as a whole, there is less heeling down on the up-stroke—hence less thrust—than there is on the down-stroke."

Gibbs-Smith thus disagreed with other bird-watchers, like B. Baden Powell, who in 1894 observed that a bird, seen from below, appears alternately to extend and flex his wings, "and it is therefore probable that the wing during the down stroke is well spread out to take advantage of the greatest area for compressing the air beneath, while during the up stroke the wing is flexed."

He concluded: "It seems extraordinary that, after all the careful observations of scientists during many centuries, we cannot say that we thoroughly understand the exact action of a bird's wing during flight."

Simon Newcomb, the 19th Century mathematician, observed: "The desire to fly like a bird is inborn in our race. We can no more be expected to abandon the idea than the ancient mathematician could be expected to give up the problem of squaring the circle!"

Professor Joseph Le Conte, the eminent naturalist, said frankly: "I am one of those who thinks that a flying machine is impossible, in spite of testimony of the birds!"

This was the attitude held by most of the scientific community before the advent of the tunnel men, who would point the way toward conquest of the sky through laboratory observation and analysis of the winds.

In California, John J. Montgomery almost stumbled onto the secret of aerodynamic lift, with his crude wind tunnel observations of dust particles blowing over cambered airfoils. Montgomery mistook their curving path for the crest of a wave, and missed the whole point—the lifting force of the lower pressure area above the wing, identified mathematically by Daniel Bernoulli.

The world's first wind tunnel was built by Francis Herbert Wenham, founder of the Aeronautical Society of Great Britain. With straightforward British logic, the Society in 1871 raised a subscription for Wenham to undertake systematic aerodynamic experiments to obtain "data on which a true science of aeronautics can be founded."

Wenham's tunnel, which he described in a paper read at the Chicago Conference of 1893, was a wooden trunk 18 inches square and 10 feet long, through which a steam-powered fan blew a 40-mph wind. Wenham went ahead blindly, experimenting with flat plates, inclined at angles from 15 to 60 degrees to the airflow. Despite their crudeness and unexpected turbulence, these experiments were encouraging. Wenham found that lift, at small angles of incidence, exceeded drag by a much greater extent than had previously been suspected.

Recognizing the need for varying the speed of the air through his tunnel, Wenham told the Society he hoped to build a larger apparatus with a wind velocity "from a gentle breeze up to a tornado that could rip the clothes off your back, or blow you away like a feather; but no flying man should mind this effect," he added.

In 1900, then 76, Wenham actually succeeded in building a tunnel that produced a 25-mph wind with a hand-driven fan spun at

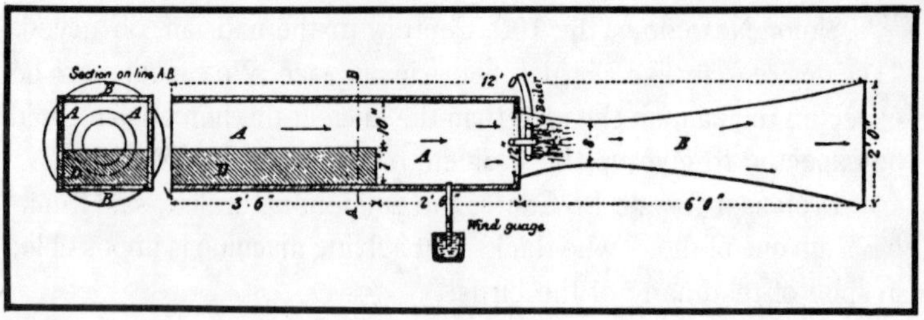

Horatio Phillips' 1884 wind tunnel was first to use narrow-throat Venturi tube design.

1700 rpm. "I could not get beyond this, as it absorbed all my strength to work it," he complained.

The first tunnel man to identify the area of lower pressure formed above a cambered airfoil was Horatio Phillips, using a homemade wind tunnel built in 1884. It was a novel device which produced a controlled air current by means of a steam jet that eliminated much of the turbulence Wenham encountered.

Far more sophisticated than Wenham's, Phillips' tunnel employed a little known principle discovered by an Italian physicist, G. B. Venturi (1742-1822) while studying the flow of fluids through narrowed channels. This was the first "Venturi tube" which Phillips described as "an expanding delivery tube of sheet iron" into which steam was sprayed through fine nozzles under a pressure of 70 psi. Rushing through the throat of the Venturi tube, the mixture of steam and air created a suction that gave a wind velocity of 60 feet per second.

In his experiments, Phillips developed his famous series of patented, cambered airfoils, some of which featured a drooping leading edge, later adopted by Louis Bleriot and other pioneer aircraft designers in 1908-1910. In May, 1893, Phillips built and successfully flew a large, powered model craft that resembled a Venetian blind, with twenty long, narrow, cambered airfoils stacked one above the other. Mounted on a tricycle gear and powered with a steam engine, it reached a speed of 40 mph flying in sweeping circles, at the end of a control line.

In patenting one of his airfoils, in 1891, which he used in the Venetian blind machine, Phillips first described his discovery of low-pressure lift: "The particles of air struck by the convex upper surface are deflected upward...thereby causing a partial vacuum over the greater portion of the upper surface..."

First to run wind tunnel measurements of pressure distribution over a wing was H. C. Vogt, of Copenhagen, in 1894. A marine engineer, he was inspired not by bird flight but by the mysterious action of the wind blowing over the spreading sails of yachts plying the North Sea. Vogt discovered that it was this action of the wind which gave the sails their graceful curves, but he was puzzled as to why the sails obtained much of their forward thrust from the leeward side, where a low-pressure area formed.

To learn more about this phenomenon, Vogt built an ingenious wind tunnel powered by the strong updraft inside the 100-foot tall smokestack of the Copenhagen Gas Works. With the help of the firm's director, Johan Irminger, he knocked a hole in one side of the chimney and inserted one end of his 40-inch tunnel.

Inside this tunnel Vogt inserted hollow airfoils that had been drilled with a number of small holes. These were opened one at a time, and the resulting pressure readings were taken from a common water gauge connected to the inside of the airfoil.

Vogt is credited with first discovering that the region of greatest lift along a wing occurs toward the leading edge where the upward deflection of air is more pronounced.

Dr. Albert F. Zahm, the scientist who helped Chanute organize the 1893 Chicago Conference on Aerial Navigation, in 1901 built history's first complete wind tunnel laboratory on the grounds of the Catholic University of America. It was financed by a New York inventor of a giant flying boat, Hugo Mattulath, who had hired Dr. Zahm as a consultant.

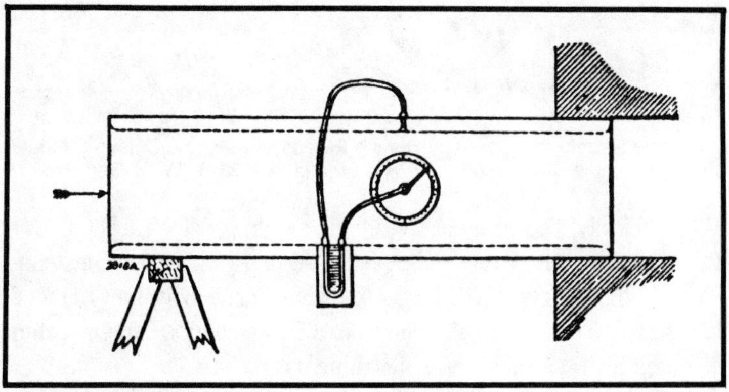

H. C. Vogt and Johan Irminger built wind tunnel inside a factory smokestack at Copenhagen Gas Works.

Multiplane aircraft, thought of by Horatio Phillips in 1890 and Octave Chanute about same time, were used in MPA designed in 1923 by W. F. Gerhardt.

Ironically, the true nature of induced drag, which accounts for the major portion of total drag, escaped him—his researches convinced him that skin friction was to blame. Inside a 40-foot tunnel 6 feet square blew a 27-mph wind from a 5-foot suction fan. Dr. Zahm invented his own measuring instruments.

His investigative methods were unusual, to say the least. Once he hung a 16-foot plank in the tunnel endwise, then sawed off one foot at a time, checking the drag reduction as he went. He did not doubt that skin friction was heavily involved.

To measure accurately the speed of his manmade wind, he floated a toy balloon down the tunnel. On its way it interrupted two thin pencils of light focussed on a moving plate of film. It worked well.

In 1902 Dr. Zahm's tunnel work was interrupted by the death of Mattulath, who suffered a stroke. The flying boat project was abandoned, but grants from the Smithsonian and Carnegie Institutions permitted him to resume his work. In February, 1904, he revealed his conclusions about viscosity, or skin friction, in a paper on "Atmospheric Friction" which he read before the Philosophical Society of Washington.

Dr. Zahm was to occupy a niche in the hall of aviation fame because of his intense laboratory investigations. He is remembered today as America's Father of Flight Theory. Not one to hog the spotlight, Dr. Zahm permitted other researchers to use his wind tunnel for their own studies. Chanute once sent him a stuffed buzzard, to check its lift and drag coefficients, and a rocketship designer, Emile Berliner, also used the facility.

Among other early wind tunnels was one built by Dr. Ernst Mach of Vienna, in 1893. First to use a wind tunnel to photograph air flow, through a glass window, Dr. Mach used a piece of wire mesh over the opening to align the current, sucked through the tunnel by a centrifugal fan at a rate of 10 meters per second. Visualization was achieved with silken threads, cigarette smoke, and particles of glowing iron. Streams of heated air, invisible to the eye, also were recorded on a photographic plate. Dr. Mach's name is used today to describe the ratio of a flow velocity to the speed of sound—called the Mach number.

Gustave Eiffel, the French engineer best remembered for the 1,000-foot tower in Paris which bears his name, and for the Statue of Liberty in New York, devoted his later years to studying aerodynamics of aircraft. He built his first tunnel on the Champ de Mars in 1909, and in 1911 built a larger one at Auteuil, later donating it to the French Government. Eiffel developed a series of airfoils which also bear his name, and which were widely used in early powered aircraft.

Two other early wind tunnels deserve mention for the contributions of their designers to a fuller understanding of the true nature of aerodynamics—Nikolai E. Zhukovski in Russia, and Ludwig Prandtl at Gottingen in Germany.

Zhukovski's tunnel, 40 feet long and 4 feet in diameter, was built in 1905 at the Koutschino Aerodynamic Laboratory near Moscow and was financed by a Czarist capitalist, M. D. Riabouchinsky. A low-velocity tunnel (14.5 mph), it was used by Zhukovski for pioneer studies of the thrust of helicopter blades, with wool tufts glued to them for flow visualization.

Zkukovski and a German scientist, M. Wilhelm Kutta, came independently to the idea of the circulation hypothesis of lift and produced the mathematical theory that for the first time fully explained how a wing flies. Their work, however, was preceded by that of Frederick W. Lanchester in England, who is credited with first suggesting the idea of total circulation of an airfoil of infinite span, and the vortex theory of lift of a wing of finite span.

Many researchers today credit Prandtl's mathematical formulation of the circulation theory of lift as one of the most useful contributions ever made in the field of aeronautics. It was Prandtl who also

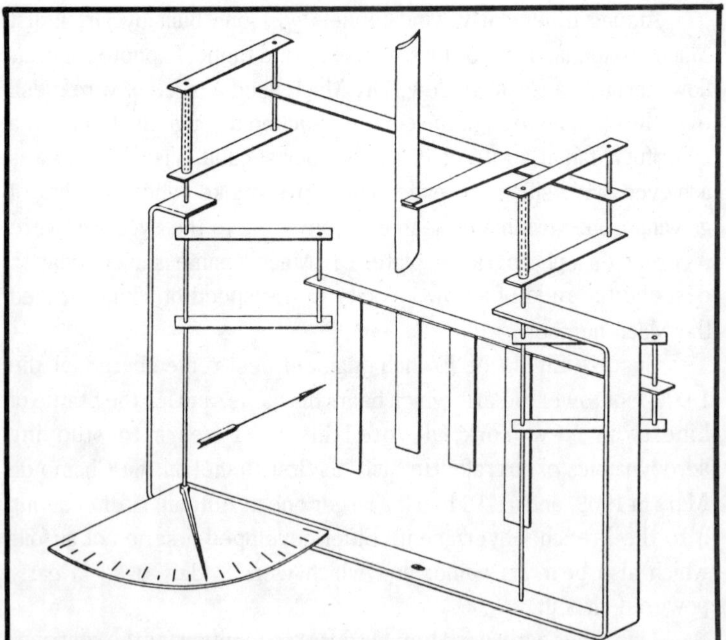

Mechanism of measuring apparatus designed for wind tunnel studies of airfoils by the Wright brothers.

first recognized that the total resistance of an aircraft is composed of three parts—skin friction (viscosity), form or pressure drag, and that produced by lift (induced drag).

Many airfoils developed by Prandtl at Gottingen are in use today, and are known as Gottingen curves.

Attention must be paid to the tunnel experiments of Orville and Wilbur Wright in October, 1901. About 6 feet long with an interior of 16 square inches, their tunnel had a glass top. A blower fan forced air through honeycomb windstraighteners at 40 feet per second.

The Wrights tested some 200 wing models of sheet metal, each set at 14 different angles of attack from 0 to 45 degrees. Other tests were run on various aspect ratio wings, and on superposed surfaces. The curved airfoils were balanced against plane surfaces, to provide a series of aerodynamic tables, after the Dayton brothers found that Lilienthal's tables were highly inaccurate. Their tunnel work was a chief factor leading to their success at Kitty Hawk in achieving powered flight in 1903.

Chapter 8
You'll Never Get Off the Ground

"I said it to Orville, and I said it to Wilbur—"

T. W. Mather, writing in *The Popular Science Monthly* in November, 1885, came to the conclusion that, despite the examples of the birds, inventors of flying machines "as a rule belong rather in a lower class" than balloonists.

"The perpetual-motion man," he sneered, "is likely also to know just how to make a successful flying machine. The pathway has been strewn with wrecks, and I fear there is a feeling that, after all, it leads nowhere in particular, unless it be to the almshouse or lunatic asylum."

But still, he added, "there are the birds."

"One cannot help thinking," Mather concluded, "that we are about as likely to make a practicable flying-machine by copying the motions of the bird as we are to make a steam road-wagon by imitating the action of a horse."

If history proved him right, the dream of imitating bird and/or insect flight is ageless and continues today, although man-powered ornithopters still appear to be beyond the reach of aircraft designers and builders.

"Nature cannot always be trusted," Mather warned. "Why should we try to copy the motions of a bird's wing any more than those of a fish's tail? Our machine, in imitating them, would be complex and liable to get out of order. The desired results can probably be obtained in a much more simple and effective way. There has been a strong current of misguided thought and invention, particularly to be noticed in our Patent Office reports."

Truly, flying machine inventors were considered rather mad in the last century. Reuben Jasper Spalding nearly lost his life when drunken cowhands attacked him, when he attempted to fly on his patented wings. Scorn, or at least amusement, greeted others who succeeded in patenting their updated da Vinci devices, such as Watson F. Quinby's batlike Flying Apparatus of 1872.

Dr. Watson Fell Quinby's secret interest in flying was typical of ingenious American back-country folk, who shunned public attention because of the ridicule of "authorities" like T. W. Mather, who lumped them with "perpetual-motion men." An adventurous, mechanically-minded descendant of English Quakers who migrated to America in the 17th century, Dr. Quinby was born in 1826 in Mill Creek Hundred, near Brandywine Springs, Delaware.

As an admirer of Oliver Evans, a pioneer investigator of steam power who lived nearby, Dr. Quinby's first attempt to design a flying machine was to be a steam-powered aircraft which also could be manually-powered. Patented in 1861, it featured screw propellers at the stern and above, with wings consisting of curious oiled-silk valves that opened on the upstroke and closed on the downstroke, reminiscent of one of da Vinci's early designs. Lack of a suitable steam engine caused him to put the idea aside.

A country doctor by profession, Dr. Quinby had gone to Mobile, Alabama, to practice medicine, but in 1849 was lured to the California gold fields. He returned to Delaware in 1852, married, and set up a new practice. In 1853 he was inspired to develop his steam-powered aircraft when the showman P. T. Barnum posted a $5,000 prize for the first person to fly across the Atlantic to Europe.

Although his 1861 flying machine was a failure, the idea of flying stayed with him through the better years of the Civil War, and in 1867, when he had more time for his own projects, he patented a new "Improved Flying Apparatus," this one a truly man-powered aircraft design. It consisted of oiled silk wings and a tail, strapped to the operator's back in a manner to keep the arms and legs free.

"In using this invention," he explained, "a motion is given to the arms and legs almost precisely like that in swimming, and the effect is nearly the same, the difference being in the density between water and air being compensated for by the greater extent of surface

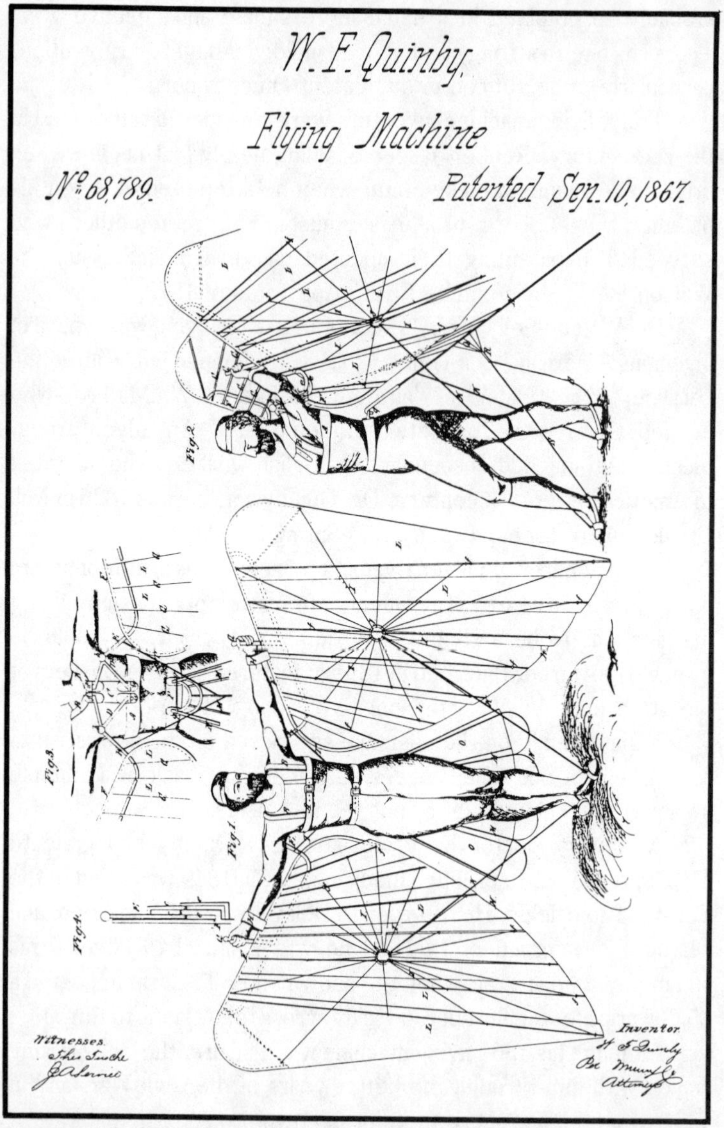

In 1869, with the war over, Dr. Watson Fell Quinby designed and patented winged MPA for operator to use by swimming motions.

presented by the wings. The wings are brought up to their full height, and then suddenly pushed forward, and by a pronation of the hands thrown outward, and then brought down by a full force or power of both arms and legs. By this method, every muscle of the body is brought into action to the best advantage."

The Scientific American, reporting on the device, felt that the use of all body muscles really *was* an improvement on earlier Darius Green machines, but warned: "We hardly think he will be able to compete with the swallows in this harness, and would advise him to start from some low point at first, so that, if he should fall down, it would not hurt him much."

Dr. Quinby apparently took the well-meant advice, according to an elderly Wilmingtonian who in 1927 recalled an attempt the doctor made to fly from the roof of a low farm building, where he had secretly built his device:

"Dr. Quinby expected in the hop-off from his elevation to get his machine 'carrying on' before he alighted on the ground fifteen feet below. But being so close to the ground, Dr. Quinby and his flying machine had a rough fall, violent enough to lay the physician up for a day or two, while his machine, badly damaged by the drop, was hidden away and never again given the opportunity to demonstrate that it could be made to fly."

For a time the good doctor, stung by ridicule, forgot flying and devoted his spare hours to writing pamphlets, one presenting arguments for and against adopting the metric system in this country: *"The Yard or the Metre, Which Will Ye Choose?"*

The lure of the sky was too strong, however, and in 1869 he patented still another flying machine, this one a set of wings whose spars served as oars, with a foot-actuated tail surface behind. Abandoning that design, in 1872 Dr. Quinby patented a new set of wings which, like da Vinci's favorite design, resembled those of a bat.

"The weight of the whole machine need not exceed 15 pounds," he said. "It is constructed inside a semicircle, all the points touching the periphery. I describe a semicircle on the floor 10 or 12 feet in diameter, and bisect this figure with a radius."

In operation, he wrote, "It is intended to start from the ground." (no more rooftop launches!) "One foot is disengaged from the stirrup, when, by raising the other foot and pushing the hands upward and forward, as in swimming, the wings are raised. Then by suddenly depressing the wings, by means of the elevated leg, the body is elevated. This alternate elevation and depression of the wings is continued as long as flight is desired. After rising from the ground the other foot may be inserted in its stirrup and both legs used."

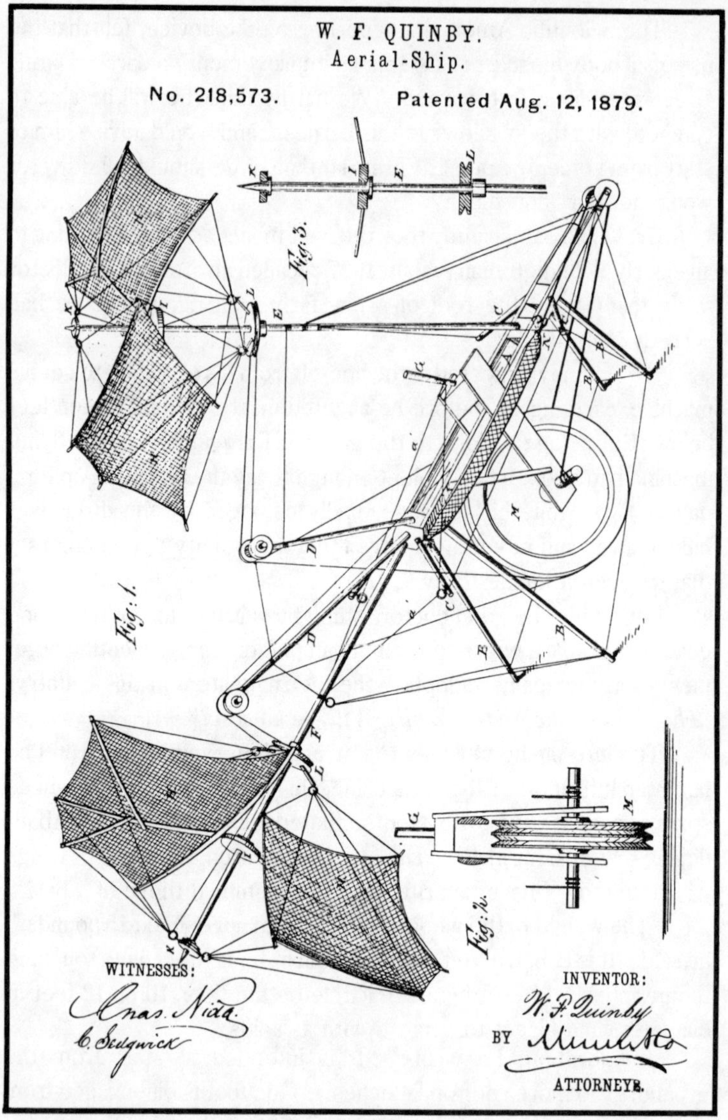

In 1879 Dr. Quinby's last patented Aerial Ship used sailwings for lift and propulsion.

In a last fling at flying, Dr. Quinby in 1879 patented his final "Aerial-Ship", returning to his original idea of providing a boat to ride in, and sets of winged sails that served for both lift and propulsion, adjustable in pitch, a crude forerunner of the modern variable pitch propeller.

After a full life, Dr. Quinby died in 1918 at the age of 92, having lived to see the conquest of the sky become an actuality, albeit with rigid-wing aircraft, not ornithopters.

Indeed, the success of the Wright Brothers in flying a rigid wing machine at Kitty Hawk in 1903 in no way deterred dreamers from their pursuit of the age-old goal of imitating bird flight with mechanical wings. Even today, serious study is again being given to the problem, from both the aerodynamic and mechanical viewpoint.

Far ahead of his time in this respect was Professor Harry LaVerne Twining, a physics instructor at Los Angeles Polytechnic High School, who in the early 1900's studied the action of a buzzard's wing in an unusual manner—he waved the wing of a dead buzzard, in imitation of its natural vibration, close to a piece of silken cloth, 2 by 3 feet, hanging from a support.

"Whenever the concave or convex surface of the wing was brought close to the cloth," he said, "it was attracted to the wing and stuck to it as though glued there."

He traced the airflow pattern in this crude fashion and concluded that "the air comes in toward a vibrating wing from above, from below, from the whole front, and from the rear for two-thirds of its length. This air is gathered into two streams, one above the wing and one below the wing, and is discharged at the rear tip of the wing for one-third of its distance. At this place the air is driven violently away from the wing, backward and downward, forming a large cone of moving air with the apex at one-third of the rear edge of the wing."

Twining, who thus came remarkably close to grasping the hypothesis of total circulation and wingtip vortex flow, set about building a set of canvas wings, went alone into the hills, and tried to fly with them. A gust of wind caught him up and roughly tossed him into a mesquite bush. He decided the matter needed more careful thought.

"The flying of soaring of birds," he concluded, "can be explained upon well-known mechanical principles, and it is surprising that these feats of winged creatures have remained so much a mystery." He pointed out that the resultant of all forces acting on a bird's wing, whether moving upward or downward, is forward in the plane of the wing. It was his belief that "the machine is held firmly between two planes of resistance, between which it glides smoothly forward."

After publishing these conclusions, Professor Twining turned for a while from flying machines to a far more subtle mystery of nature—the weight of a mouse's soul. In 1907 he busied himself in his high school laboratory, weighing mice just before and after killing them. He repeated his experiments on grasshoppers and kittens.

It was his snap conclusion that an apparent loss of weight before and after expiration proved the existence of a soul, but later he retracted that view—he concluded the weight loss was due to the loss of body gases and moisture, not ectoplasm.

Two years later Professor Twining was back at flying machine work, as secretary of the Aero Club of California, which was sponsoring an air meet at Dominquez Field near Long Beach. He revealed in 1909 what he had been up to: "Some five years ago," he reported, "I made a machine with wings and took it into the hills of San Pedro, in the darkness of early morning hours. There for two days, like a condor with a full stomach, I tried to get off the ground, but it would not work—the first machine that a man makes never does. But I learned a great deal by experiment, and with the added knowledge I propose in the near future to build another machine. My first machine weighed 35 pounds. The wings had a spread of 24 feet from tip to tip and a width of 2 ½ feet."

Professor Twining spent the summer of 1909 working on his second wing flapper, whose mechanism he described in a patent, issued to the inventor August 20, 1912. Completed in the fall of 1909, the machine was wheeled from Twining's workshop for its initial test hop. The appearance of the ornithopter in the streets of Los Angeles prompted considerable interest. A story in the *Los Angeles Daily Times* on August 13 reported:

HIS FLIP-FLOP
TWINING'S BIRD
ABOUT TO FLY

Professor Believes He Has
The Real Airship

"Professor H. LaV. Twining...has almost completed a machine with which he believes he will fly. In the workshop of J. H. Klassen is a frail aircraft, equipped with two great wings that suggest some monstrous bird of prehistoric time. The wings, carefully modeled after those of an eagle, measure 27 feet from tip to tip, the operator's station being three feet wide.

"Professor Twining does not intend to rise far from the ground at first, and is confident that when it does get into the air there will be

Professor Harry LaVerne Twining of Los Angeles built an ornithopter reporters called the "Flip-Flop." It barely got off the ground.

"His ground runs were one of the highlights of amusement," says Lippisch. "Oskar Ursinus (who later formed the Muskelflug-Institut, or Institute of Man-Powered Flight), running in front with a handkerchief to determine the wind direction and Brustmann high on the wheels flapping the wings like a newly hatched chicken."

Dr. Brustmann's machine was a flop, but he had something more important to contribute to the MPA program. As a sports physician who trained athletes for Olympic competitions, he had experimented on how to reach a man's highest power output, although his figures were quite optimistic.

What intigued Lippisch was the trend of Dr. Brustmann's figures, which showed the relation between human-power output and its duration to be a kind of hyperbolic function. "This relation," he recalls, "showed that the human body stored a certain amount of energy which could be expended over a short period, and high-power output could be expended for a period of less than one minute. At least, I knew that for a short time it seemed actually possible to fly by one's own power, since a lightly built glider plus man would not need more than one horsepower to be flown horizontally."

The upshot was that Lippisch decided to build a very light test craft, as a flapping-wing MPA. "The reasoning was based primarily on the physiology of the function of muscles designed for intermittent use," says Lippisch. Besides, he felt that wing flapping would have better efficiency and would cause less additional structural weight for the required power-delivery mechanism. He still held this view in 1960. Dr. Brustmann suggested utilizing arm and chest muscles in addition to the leg muscles for propulsion power.

Lippisch's man-powered ornithopter was built by Alex Schleicher in Poppenhausen as a high-winger with an open pilot seat and a covered fuselage behind it. The wings were moved by leg action, as in rowing from a sliding oarsman's seat.

Each wing was supported by a triangular strut, the end of which slid along a guide rail on the fuselage, controlled by cables. Shock-cord rubber tension springs in the rear fuselage were adjusted to provide wing dihedral in normal gliding flight.

The wing was built around a single spar, with thin veneer on the leading edge and fabric behind. The root section was highly cambered, and the tip almost symmetrical. The craft's empty weight was around 50 kg, or 110 pounds, higher than anticipated, due to the wing-flapping mechanism.

Professor Harry LaVerne Twining's ornithopter of 1909-10.

no danger...He intends to conduct his first experiments in private, and will begin today to endeavor to operate the machine."

Many years later, Professor Twining, interviewed by the author, recalled what happened: "I flapped the wings furiously, but was only able to get the front of the machine off the ground. I was on the point of making alterations in the machine's design, when it became apparent that the rigid wing type aircraft was vastly superior for any commercial purposes. I still believe, however, that my ornithopter can be made to fly with only slight modifications to the original design."

As President of the Aero Club of California, Professor Twining on January 10, 1910, assisted in getting America's first aviation meet rolling. The Professor's Flip-Flop, as the newspapers dubbed his ornithopter, went on exhibition, but no attempt was made to fly it.

That summer Professor Twining, with the help of his physics students, completed construction of a conventional monoplane. It did manage to get into the air for a few feet, but its Model-T engine was underpowered.

A noted German sailplane designer, Alexander Lippisch, holds the honor of developing the world's first ornithopter man-powered aircraft to perform with any success, in the year 1929. Lippisch first came into prominence in the post-World War I period from 1920 to the mid-thirties, as Technical Director of the Rhoen Rossiten Gesellschaft at the Wasserkuppe. Among his outstanding designs was the *Professor*, *Wien*, *Falke*, and *Fafnir I* and *II* sailplanes. Work on a tailless Delta-wing glider led to his design of the Me-163 rocket interceptor, the Komet, in which Heini Dittman becave the first man to fly at 1,000 km/h (620 mph) in level flight.

Lippisch later was spirited to the United States, at the end of World War II, and continued his man-powered aircraft researches well into the 1960s. At that time, he recalled in the Journal of the Royal Aeronautical Society his early MPA work on ornithopters.

During the 1920s at the Wasserkuppe, a time when he was designing airfoils and calculating induced drag according to Prandtl's newly discovered wing theory, he was frequently called upon to evaluate MPA's built by others. Seen from a distance, he recalled, they "resembled more one of those prehistoric bird petrifacts." He called them *Pteroptodons*.

Lippisch felt it would be helpful and inspirational to invite these inventors to attend the soaring competitions at the Wasserkuppe, and among them was a medical doctor, Martin Brustmann, who arrived with a kind of ornithopter constructed around and above a bicycle.

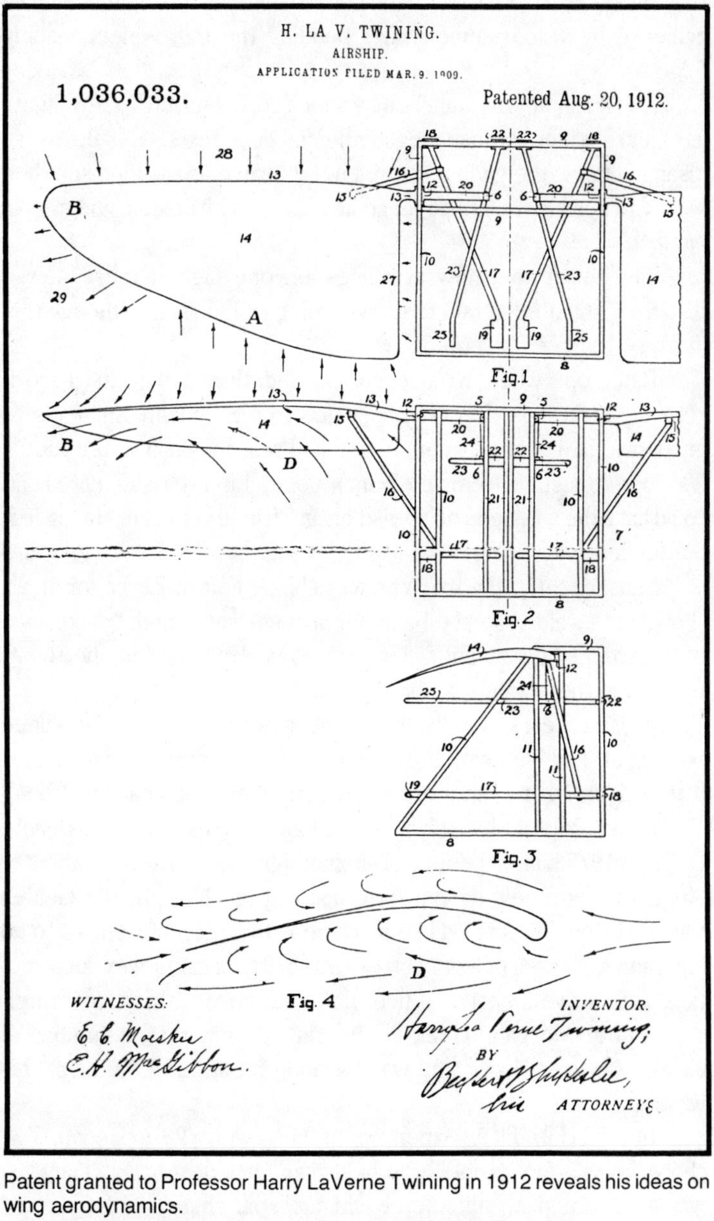

Patent granted to Professor Harry LaVerne Twining in 1912 reveals his ideas on wing aerodynamics.

A young athlete, Hans Werner Krause, was selected to test fly the machine, early in 1929. A single control stick operated both rudder and elevator. There were no ailerons, because Lippisch believed lateral control could be obtained by flapping one wing harder than the other. Wing twist on the downstroke, he believed, would be achieved by aerodynamic forces bending the thin veneer leading edge.

The machine was launched by shock cord, and after a few flights as a pure glider Krause began the flapping tests. Results were disappointing—the glide angle did not improve at all. Lippisch then went to a flexible trailing edge, adding Taube-like flexible wingtips of bamboo.

The result was now startling—on one flight Krause flew a distance estimated at 600 feet, according to Lippisch, although this distance has been disputed.

Later on, when Krause complained that it was hard work pumping the wings, Lippisch appealed to his romantic instincts to encourage him to try harder. If he could reach a small water puddle 300 meters distant from the launch point, Lippisch suggested, he could have the weekend off to visit his girl friend in Berlin. He made it on the first try.

Subsequently, the machine was shipped off to Berlin for an air show, and a welded steel tube undercarriage was added, "making it a truly earthbound machine," Lippisch says. Krause died shortly afterward in the crash of a sport plane.

In later years, Lippisch felt the machine could be modified somewhat, and built with modern lightweight materials in a manner that could make it even more successful than it was back in 1929.

An admirer of Lippisch's work, Douglas Kruse, of Long Beach, Calif., in 1977 built a model of a similar high-aspect ratio ornithopter (40:1) to test a new design for a flapping mechanism, the Golden Eagle. His goal was to build a machine with a span of from 74 to 90 feet, and wing area of from 160 to 190 sq/ft., all cantilever construction, with a metal spar and fiberglass and polyurethane covering.

And in Morgan, Utah, Klaus Hill, designer of a number of ultralight powered aircraft, was working on an ornithopter at this writing.

In 1973 Hill built an experimental MPA with a 28-foot span, 5′ 3″ chord, and a 5-foot propeller to be driven by pedal-power. The wing was built around an aluminum U-channel spar shaped with foam and

Klaus Hill of Morgan, Utah, designed all-metal MPA for ground-run evaluation of design, in effort to win the Kremer Award.

fabric covered. Fuselage was beautifully worked sheet metal, highly polished. The airfoil was symmetrical, of 12 percent thickness, with a drooping leading edge. It weighed 90 pounds empty.

"It was strictly an experimental craft," says Hill. "I built it to run taxi tests to check certain aerodynamic and mechanical design features, hoping to go from there toward winning the Kremer Award."

Other inventors around the country still hold to the idea of flying with flapping wings. In Inglewood, Calif., Rex Pagett went to the concept of using the powerful back muscles to power his wings, which attached to the shoulders of the pilot.

But if any wing-flapper finally succeeds in achieving controlled flight over a period of time, it might well be the Orni-Plane, a design of Percival H. Spencer, of Sun Valley, Calif. Designer of the Republic SeaBee, the Spencer Amphibian Air Car, and numerous other aircraft, Spencer in 1977 celebrated his 80th birthday flight testing a thing he calls the Hydro-Glider.

Spencer also made a small fortune with a patented commercial toy called the Wham-O Bird—625,000 were sold in the late 1950s. Rubber-band powered, it flaps its way around the sky with the greatest of ease. His patented toy features a link between the wing flap and the tail feathers, to permit it to glide properly regardless of the wing position when the power runs down.

Earlier in his career, Spencer had designed numerous other ornithopter models, but his most successful one, he claims, is an 8-foot wingspan model of the Orni-Plane that hangs in his bedroom.

His dream is to secure financing to build a man-sized Orni-Plane, fly it once, then retire it to the Smithsonian Institution.

One other machine bears attention here—a patented flapper called the Orniplane, developed by three Russian inventors, I. N.

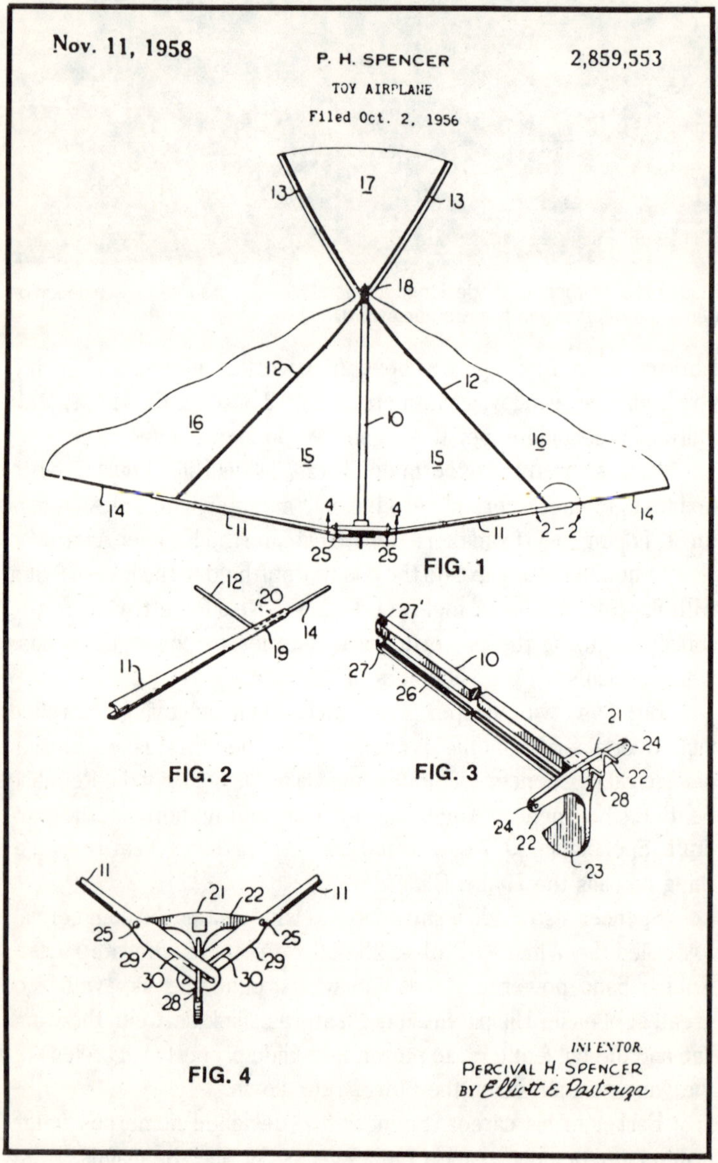

P.H. Spencer in 1958 patented a Toy Airplane that flew well as a wing-flapper. It was sold commercially as the Wham-O Bird.

P. H. Spencer's Wham-O Bird in flight.

and R. I. Vinogradov and V. M. Andreyev, whose secret lies in its feathered wingtips. The patent description explains:

"For the purpose of increasing thrust and lift, at the wingtips are used propelling tips (manus) consisting of feather-like vanes, made of synthetic material, with metal longerons. The vanes are arranged in fan-shape in one block executing reversive-torsional oscillations, transmitted by a helicoidal automat, consisting of hinge-jointed telescopic wing section longerons and power cylinders with supports on telescopic longerons and spring-blocking in the hoops of the cylinders."

Their machine also included a boundary layer control system, and a hydraulically-operated flywheel system (elastic-plastic) for automatically changing the wing profile. At least, that's what a translator at the Air Force Systems Command made of it.

Any why, after all, would anybody really want to fly like a bird, with flapping wings? "The best reason is simply that it hasn't been done successfully yet," says Walter H. Carnahan, who in 1975 conducted a forum on ornithopters at the Experimental Aircraft Association's summer convention at Oshkosh, Wisc.

Carnahan adds: "Have we forgotten that flying can also be fun? The birds haven't forgotten. On a day that is so gusty that old, bold human pilots are sitting on the ground reading about flying, the seagulls take to the air and cruise in turbulence, having a ball!"

Chapter 9

Bugsville

Two hundred sixty million years ago, in the Paleozoic epoch of the world's history, there lived a species of gigantic dragonflies called Protodonata. They flew on two pairs of graceful wings, with a span of two feet. The wings were formed of tough membranes stretched over an elaborate framework of veins.

The Protodonta were strange creatures with a thousand eyes, who pursued their loved ones relentlessly and mated in mid-flight. They could fly from dawn til dusk on warm, sunny days—hovering, darting forward or backward, with amazing precision.

Today, scientists marvel at the fossilized remains of these fantastic flyers, preserved in the Carboniferous strata at Commentry, France. They were the largest of some 4,870 species of Odonata, commonly called the Dragonfly. This is not a large figure, considering that the total number of living forms of insects runs between 2,000,000 and 4,000,000, dominating by sheer number all other forms of life. Of this huge number, roughly 750,000 species can fly.

One man who became fascinated by the darting flight of a dragonfly, catching mosquitos on the wing, hovering, backing up or rising vertically, was William Bushnell Stout, the noted designer of the Ford Trimotor all-metal transport. Once asked whether he was a descendant of David Bushnell, designer of a famous Civil War sub-

William Bushnell Stout, designer of Ford Tri-Motor transport, built experimental models of mechanical dragonflies.

marine, he shook his head and smiled: "All the world knows he died a bachelor!"

Stout was a man of typical Yankee ingenuity. Among his contributions, besides the Ford Trimotor, forerunner of toay's airliners, were the Sky Car, a family flivver plane with a tricycle landing gear; the first light, streamlined railroad car, and the 1932 Scarab, a streamlined, rear-engined autombile of the future.

Hence, his interest in the flight of the dragonfly was no passing fancy. He put other things aside and researched the subject at the

Congressional Library in Washington, D. C. There was not a single reference to the mechanics of insect flight!

Next Stout looked up Professor Leigh Chadwick, who had taken movies of insects flapping while held by forceps, using a stroboscopic light. Dragonflies, he discovered, flapped their wings 30 times a second, while common house flies and bumble bees flapped at more than 300 beats a second!

He recalled reading in Scientific American, in 1884, that a scientist named Paul Mazo, in Lausanne, Switzerland, had calculated that a fly's wings beat at 330 strokes a second, and mosquitos at 2,000 strokes a second. "Therefore," Mazo had concluded, "a number of common flies weighing 20 tons, and 130 tons of mosquitos, would require one horsepower to sustain themselves in the air."

That wasn't good enough for Bill Stout. He looked up a beekeeper, who told him that by putting a landing board at a 60 degree slope in front of the hive, instead of a flat landing board, his bees flew in with 30 per cent more honey on each flight. "The bee is flying close to a stall with his load of honey," Stout was told. "If he lands on a flat board, his hind legs land first and his front legs slap down and break off. If he lands on a slanting board, all six feet land at once; hence, 30 per cent more payload."

All this sounded like more folklore, so Stout decided to set up an experimental laboratory to learn first hand all about insect flight. Using a common music stand and a pivoted arm, he mounted an electric motor at one end and attached insect models at the other, so that the artificial wings were flapped by electric power. He called his device a "Geflopigus Giganticus".

From microscopic studies of actual insects Stout concluded their wings all had one common design feature—a stiff leading edge spar, getting more and more flexible from root to tip, with a flexible membrane behind the spar. "As the spar flaps up and down with some kind of motion," he said, "simple, harmonic, or otherwise, the trailing membrane takes various shapes and thrusts the air backward."

Consequently, Stout began building his model bug wings around a curved leading edge spar, tapering from the hinge point to the tip, where it curved around to the rear. He glued a sheet of tissue paper to the spar, so that its springing action held it in tension.

How to flap the wing was the next problem. A simple up-and-down stroke in a fanning action wouldn't do—others had tried that and failed. But on his "Geflopigator" the darned thing did work—to his surprise the crude wing flew around and around at some 20 mph.

To Stout, the whole exercise was simply one he hoped youngsters would follow up on. It was educational rather than demonstrative of any particular design for a practical "bug machine." He said: "I claim no results—only the beginnings of a new interest in the oldest and most efficient flight method of all."

In 1953, when the author was a reporter on the *Los Angeles Daily News*, he interviewed Bill Stout in his laboratory in Phoenix, Arizona. At that time, Stout firmly believed that a research and engineering group could carry on his work and come up with a one-man ornithopter you could fly to work and back, over all that freeway traffic.

He likened the action of flapping flight to that of a helicopter blade, which must change its angle of attack as it swings from the forward to the backward stroke. "In our flapping wing airplane," he

Stout launches rubber-band powered toy flapper.

said, we have to do the same thing, to get a constant lift at all times, and no bobbing up and down."

Early experiments on his whirling arm "Geflopigus" convinced him that the fly-type wing wouldn't work, as it produced more lift than thrust, and much of the wing area was simply parasite surface. Then he got the bright idea to study dragonfly wings.

"I photographed dragonfly wings to check on their structure," he explained. "There was the complex nature of the joints, the possibilities of effects of changes of blood pressure in the tubular spars, the structure of the hinge joint, and its angle at various points of flight."

Further study of the dragonfly wing convinced Stout he needed to know more about its intricate flapping action. An idea struck him—"what is the difference between whirling a single helicopter blade through a complete circle, or two blades in a half circle?"

The problem of reversing the blades every 180 degrees a couple of hundred times a minute "gives a structural man the jitters," he grinned. "However, the problem is not so bad as it seems. The two blades from opposite sides come together at the end of the stroke if you want to make a complete half circle, so that between these two wings can be built up an air pressure, an impact for an extra amount of strength at the end of each stroke, using up the inertia and taking it off at the end of each stroke. This is what you hear when birds take off suddenly in a steep climb, and the wings hit each other at the top of each stroke."

Further study of different insects revealed something important to Stout—their wings were pivoted in such a way that the trailing edge moves around the leading edge as a pivot, as the wings flap up and down. Through large arcs the wing actually acts like a helicopter blade, in effect, two full, connecting half circles of thrust.

"Then," he pointed out," after the insect is off the ground and in flight, amplitude decreases, the wingtips do more of the thrust work, the wing does more of the lift from the speed of progress, and we have a machine very efficient in horizontal speed—impossible in a helicopter."

Put another way, Stout maintained that with flapping wings you have the potential of helicopter-type takeoffs and landings, with better-than-airplane forward speeds. "Also," he explained, "full engine power and flywheel effort can be used for hovering and for vertical descents, after we learn more about what a wind does. The principle is there, and all the rest is accomplished except the engineering."

Stout built a second, improved "gefioppter" using a discarded vacuum-cleaner motor for a powerplant, and the first wing tried on it was constructed of pine, string, and tissue paper, with some liquid glue and thread. "It lasted about a minute," Stout said. He was impressed by the huge stresses involved.

Later designs were made with spars of split pine, with bamboo tips, and membranes of tissue paper, gold beaters skin, Saran, Cellophane, and other materials. The big problem was eliminating wrinkles.

Stout was astonished at the success of his new whirling arm machine and what it revealed, and immediately made plans to develop models of the dragonfly, nature's perfect flying machine.

Stout poses with mechanical dragonfly prototype MPA.

On my visit to Stout's laboratory, I witnessed one amazing flight of his mechanical dragonfly. It began with the whirring of the electric motor and the fluttering of the four wings as they beat up and down in a sort of figure-8 motion. Slowly the arm began to rotate. Stout stepped up the rheostat. Faster and faster the dragon-fly wings flapped. The noise rose to a high scream and the dragonfly model became a blur. He switched off the power, and the wings slowed to a stop.

I felt that I had looked into the future, as Stout showed me a model aircraft with a balsa-wood fuselage, with a tiny figure in the cockpit. He linked it to the rotating arm, beneath the wings, and as it began traveling in circles I believed that someday such a machine could be built and flown.

Bill Stout did not live to see a full-size model of his dragonfly machine fly, but others since have carried on his research work on insect flight, with more sophisticated equipment. At the University of Cambridge, Torkel Weis-Fogh discovered something overlooked by earlier aerodynamicists.

Reporting on his studies in *Scientific American*, Weis-Fogh explained: "The lift and propulsion of aircraft depend on the action of airfoils that move through the air steadily. In contrast, the muscle-powered flight of birds, bats, and insects depends on the flapping of wings, which introduces a degree of nonsteady airflow. Nonsteady aerodynamics is thus inherent in natural flapping flight."

Weis-Fogh recalled the work of Otto Lilienthal in 1889 in his classic analysis of the flying stork: on the downstroke, the wing's leading edge twists increasingly downward moving outward from base to tip. This twisting does two things. It adjusts the angle of attack to the direction of flight, and tilts the resulting force forward to provide both propulsion and lift. Conversely, on the upstroke the leading edge twists upward to keep the angle of attack small, providing some lift but no propulsion.

(This conclusion is in direct opposition to that of Professor Harry LaVerne Twining, who stated: "The upstroke of the wing drives the bird forward in the plane of the wing just as surely as the downstroke does.")

Weis-Fogh's conclusion was that two non-standard aerodynamic principles are involved in hovering flight—mechanisms he calls the *Clap-Fling* and the *Flip*. It is the Clap-Fling that pro-

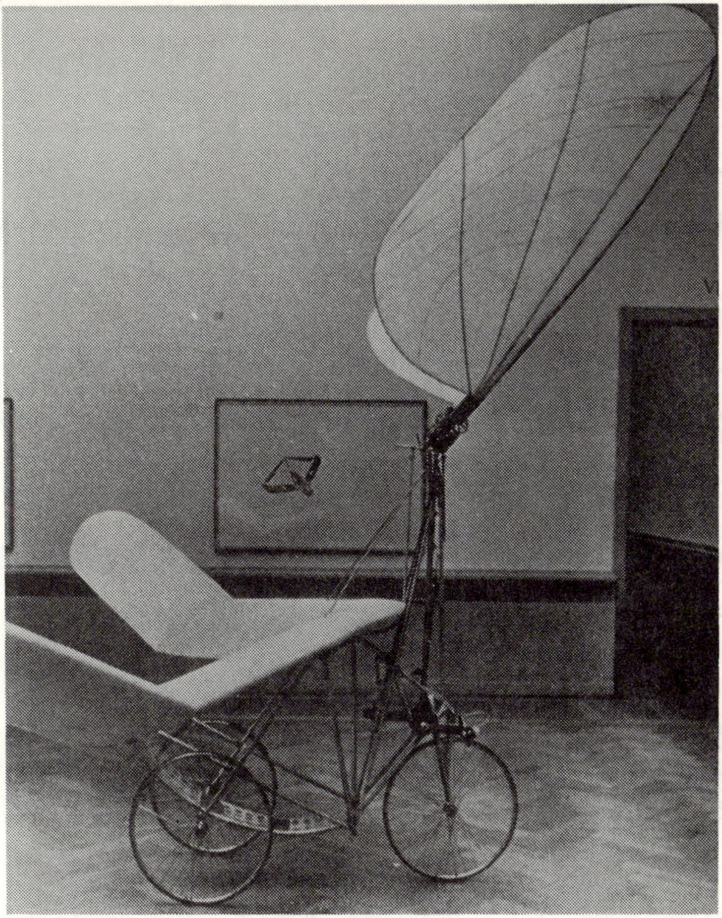

A Pop-Art MPA not designed actually to fly, built in 1977 by Belgian artist PANAMARENKO. He calls it the UMBILLY II.

duces the noise of some birds in vertical takeoff flight, when their wings clap overhead, as Stout reported.

In high-speed photography of the wasp *Encarsia formosa*, Weis-Fogh discovered that the insect's two interlocked wings on each side clap together behind its back, rest for a half millisecond, then suddenly fling open to oscillate on the downstroke to a position short of clapping underneath.

At this point the wings "flip" or twist suddenly through 120 degrees, to begin the upbeat.

By this mechanism, he explains, after the clap, when the wings open, air rushes in to fill the partial vacuum created between them, a

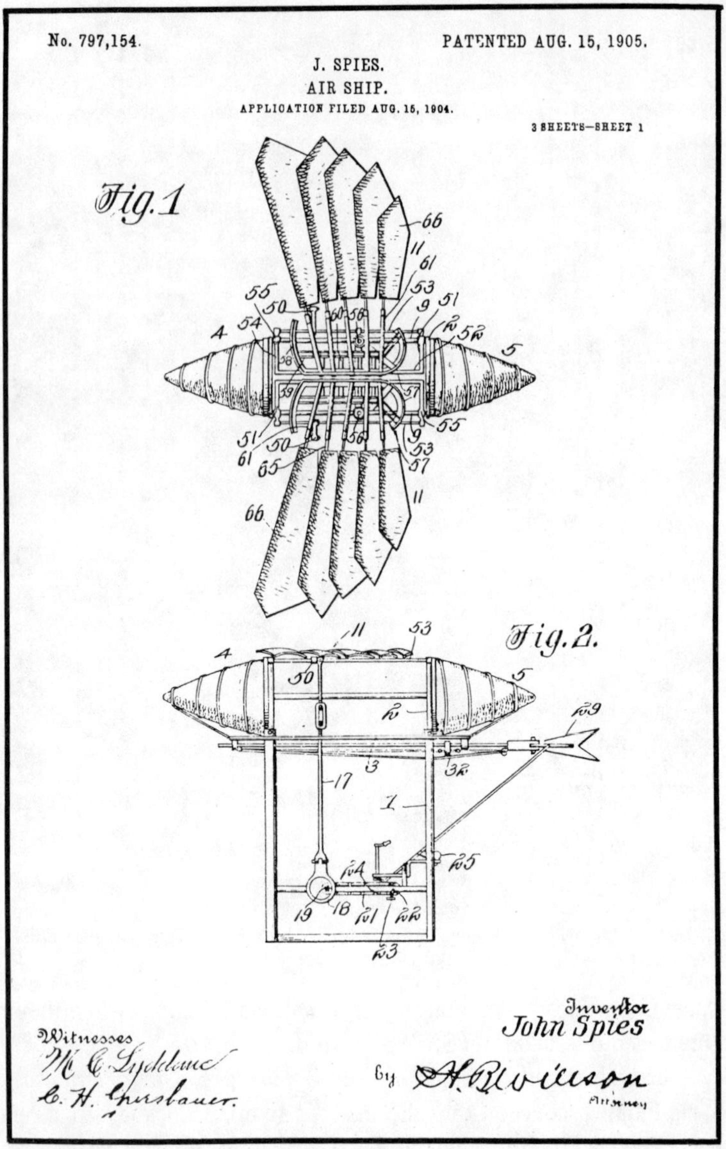

Man-powered ornithopter patented in 1905 by J. Spies.

potential flow that creates two symmetrical vortices of equal and opposite action.

Historically, says Weis-Fogh, as early as 19 B.C. Virgil observed that the rock dove claps its wings once or twice at the start of a steep climbing takeoff. Etienne-Jules Marey in France in 1890 demonstrated that the noise was produced by the wings clapping together behind its back.

Weis-Fogh admits that much research lies ahead before a full understanding of air circulation, lift, drag, and other aspects of nonsteady aerodynamic effects can be attained.

It remains doubtful, however, whether flapping-wing flight can ever be adapted to man-powered aircraft, except to extend glides, as Lippisch accomplished in 1929. P. H. Spencer, designer of the world's only really successful ornithopters, says the requirement for MPA flight by ornithopter propulsion would lie between 5 and 10 horsepower, far above the ability of the strongest athlete.

Spencer's engine-powered ornithopters, on the other hand, have been amazing. One gull-like machine, now in the Smithsonian Institution, built in 1959, with an 8-foot wing-span, performed beautifully. "It used to fly up into a clear sky so high it would disappear from sight," he says. "You could hear it flapping away, but you couldn't see it!"

A really "outa-sight" performance!

Chapter 10
The First MPA Race

Ever since Dr. Watson Fell Quinby was inspired to design a man-powered aircraft by the great showman Phineas T. Barnum, who in 1853 posted a $5,000 prize for the first man to fly across the Atlantic, the promotional value of man-flight projects has been a big factor in their development. With no obvious commercial or social value, they did offer excitement and adventure, and were sure-fire attention getters.

When Wilbur Wright made his first powered flight from the plain of Auvors, near Le Mans, in France on August 8, 1908, and in the following month amazed Europe by flying 52 miles in 1 hour 32 minutes 25 seconds for the world'd endurance record, the feats were headlined around the world. Wilbur ended his year of triumph by flying his biplane 77 miles in an amazing 2 hours 20 minutes 23 seconds, to win the 20,000-franc Michelin Trophy.

Officials of France's firm of Peugeot, makers of popular motor cars, did not miss the point—the flying craze had a tremendous advertising potential. But after the achievements of Wilbur Wright, how could they top them? The answer, they decided, lay in man-powered aircraft, which they called "Aviettes." Thus on February 1, 1912, Peugeot posted a 10,000-franc prize to stimulate interest in an MPA competition.

Flying Magazine, in its July 1912 issue, commented: "Why it should have suddenly been deemed easy to fly without a motor when it is precisely the motor, coming after thousands of years of fruitless yearning, which make flight possible, it is hard to say. But after the Peugeot firm had offered 10,000 francs for the first man who would traverse ten meters in the air without the aid of a motor, and had stated the hour at which entries would be received for the contest, twelve competitors were standing in line at that hour, apparently imbued with the idea that the first one to attempt the feat would succeed in it.

"When the entries for the first trials on June 2 closed they numbered exactly 198, and included such familiar names as Gabriel Voisin, Colliex, Ladougne, Duthiel, Poulain, de Kergarion, and Goupy; but as the various competitors practised for the event it became evident that the prize would not be so easily won after all and, seized with misgivings, the organizers offered a preliminary prize of 1,000 francs for a 'flight' of 40 inches at a minimum 'altitude' of 4 inches for 'winged' propellerless bicycles.

"When on the morning of June 2 this prize, which at first seems (and probably is) within the capability of some bicycle acrobats, it was realized that the Peugeot prize would not be won, only 23 of the 198 entires showed up and, notwithstanding countless attempts, not one of them got off the ground for the fraction of a second.

"It is evident that the problem is not being approached in the proper manner, but the experiment is nonetheless interesting and may lead to more fruitful ones along this line. There are two radically different kinds of Aviettes, the man-propelled aeroplanes in which propellers are driven by man power, and the 'winged' bicycle which has no propeller and merely depends on its speed to make the required jump.

"Two weeks ago Lavalaed, the middle-distance bicycle racer, is claimed to have 'flown' 44 inches at 8 inches elevation on his winged bicycle. From motion pictures of this feat the contrivance certainly appears to make little hops which take it completely off the ground, but there is not the slightest impression of a flying machine, and the wings seem more of a hinderance to getting up speed than a help when Lavalade makes his wheel hop."

Lavalade cheated a bit on his performance of May, 1912—although the Peugeot rules did not cover a ramp-assisted takeoff, he pedaled his machine up a short incline before plopping back to earth less than four feet distant.

Baron Fredericks's pedal-powered MPA of 1910, Germany.

Some of the Aviettes entered in the Peugeot—competition were quite novel. Mons. Didier's sported a pair of butterfly wings on the handlebars. Mons. Vincent's carried tandem fabric wings fore and aft. Mons. Malby's Bleriot-type wings were attached to a specially designed tricycle with the third wheel behind, not in front.

Didier decided to modify his flapping wings late in 1912, substituting two small front wings and a pair of surfaces over the rear wheel. With this arrangement he succeeded in winning a 2,000-franc prize, posted by the Michelin tire firm for a five-meter flight, on December 21, 1912, in Paris.

The Peugeot prize remained the big one for nine years, and in 1920 the rules were revised to include a flight in opposite directions to compensate for any favoring wind. On July 9, 1921, a favorite competitor, Gabriel Poulain, wheeled out his machine, built at the Nieuport factory, at the Longchamps race track.

It was long before sunup, and more than a hundred spectators were on hand when Poulain's "aerocycle," sporting a pair of staggered wings, made its first "flight."

As Dr. Paul MacCready had learned in later years with his Gossamer Condor, the winds were lighter and the air pressure heavier in the cool predawn hours. Poulain made several attempts, in both directions, before finally succeeding, shortly after 6 o'clock, in completing two flights at an altitude above one meter and roughly 35 feet in distance. In winning the 10,00 franc Peugeot prize, Poulain had one big advantage—he was a qualified power pilot as well as an experienced bicycle racer.

Another Peugeot prize of 20,000 francs was posted for the first Aviette to fly 50 meters, but it went unclaimed. Poulain felt that it was not sporting, because it required a flight in both directions for a considerable distance. "Do they want to destroy the spirit of the inventors and also their enterprise?" he demanded.

Detailed accounts of the Aviette competitions are given in D. A. Reay's reference: *The History of Man-Powered Flight* (Pergamon Press, 1977).

The Peugeot contests brought considerable favorable publicity to the auto makers, and inspired others throughout Europe and the rest of the world to attempt Aviette flights as a simple sport. One enthusiast, Baron Fredericks, of Germany, built an Aviette with a tractor propeller that spun quite close to his nose as he furiously pedaled his two-wheeler. A second drive chain turned the rear wheel, and lift was provided with a spread of monoplane wings and a tail.

With powered flight banned in Germany by the Treaty of Versailles after World War I, MPA competitions there in the 1920s took their place beside sailplane contests at the Wasserkuppe. Consider the situation at that time, an important moment in aviation's march of progress:

In the German heartland of Hesse, nestled between the Fulda River and the Iron Curtain, lie the historic and lovely forested Rhon Mountains, a volcanic upthrusting where slope winds blow steadily over the highest peak, 3,117 Mt. Wasserkuppe. Here the right man and the right machine appeared at the right place at the right time.

Wolfgang B. Klemperer, born in Dresden in 1893, flew with the German Air Force in World War I, and in 1920, again a civilian, graduated from Dresden's Institut of Technology to become an assistant to the distinguished aeronautical engineer Professor Theodore von Karman, with whom he worked on wind tunnel designs of light planes and soaring craft.

Forbidden by the International Allied Commission of Control to build high-powered aircraft, von Karman's group sought other means to keep Germany's highly-developed aeronautical technology alive. Klemperer was well aware of the rich potential of the Wasserkuppe for slope soaring; as early as 1910, two Darmstadt high school boys, Hans Gutermuth and Berthold Fischer, had made long glides there in Lilenthal-type machines. Gutermuth in October, 1912, stretched a glide to more than 2700 feet.

In 1920 Oskar Ursinius, editor of the German flying magazine *Flugsport*, organized the first glider meet at the Wasserkuppe,

Wolfgang Klemperer was pioneer in man-powered flight studies.

urging German youth to bring their bedrolls and their machines for a two-month campout. A total of 24 showed up. Performances were not up to expectations, however, and one youth, Eugene von Loessl, was killed when his glider broke up in severe turbulence.

Enthusiasm was dampened by the accident, and the meet would have ended then and there had not Klemperer shown up with a mystery glider, his *Black Devil*. Built by a group of students who called themselves the Aachen Aeronautical Society, the Aachen machine had thick, swept wings with a slight reverse camber and non-balanced ailerons. Wing loading was a mere .32 lb/sq. ft.

On its first flight, launched by shock cord, the *Black Devil* soared effortlessly for more than a mile—6006 feet, a new world's record. The effect of the Aachen glider's appearance was electric, and similar types were begun for the second Wasserkuppe meet of 1921. Two distinct glider types emerged. In one, the goal was to achieve a flatter glide ratio by reducing resistance to a minimum through careful streamlining. In the other, they sought to reduce the wing loading and achieve a better L/D (lift to drag) ratio and hence a better glide angle by making the wings longer and narrower.

Of 34 machines that showed up in 1921, three held special interest—Klemperer's *Blue Mouse*, a refinement of his *Black Devil*; an open primary type, the Munchen glider flown by Karl Koller; and another mystery ship, the *Vampyr*.

Oskar Ursinius in 1920s sponsored German youth glider meets that led to MPA development.

Wolfgang's historic Blue Mouse glider.

The latter craft, designed and built by members of the Hanover Institut of Technology, was a well-streamlined, high cantivered monoplane similar to the wartime Fokker F3. The pilot was almost completely enclosed in its small cockpit, and rubber footballs were used for landing shock absorbers.

Assigned to the design group on the *Vampyr* project were some of Germany's top aerodyanmicists, including Walter Blume, a former jadgstafel commander who had served on the Western Front. They could devote full time to the project, as German aircraft production was at a standstill. To Blume, this was the best thing that could have happened: "We could forget the inefficient relics of wartime," he

Famous Vampyr sailplane of the 1920s was breakthrough in aerodynamic design.

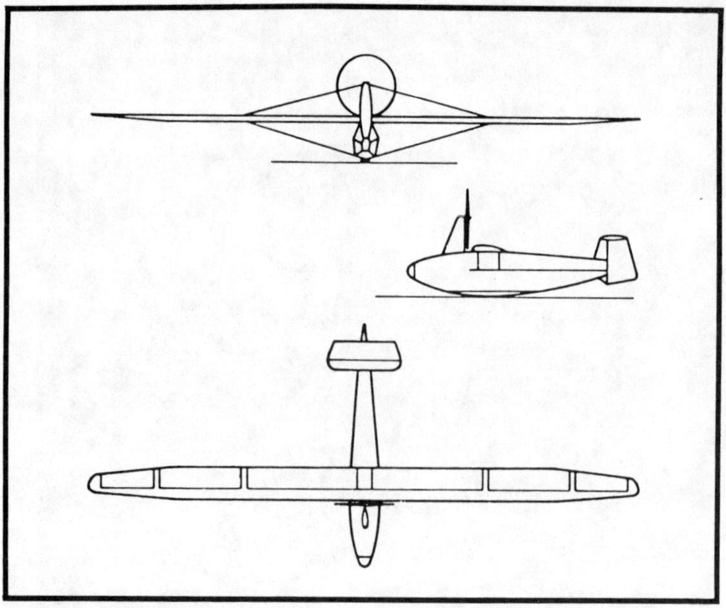

Haessler-Villinger MUFLI competed for 500-mark prize posted by German magazine Flugsport for first MPA flight of one kilometer.

said. "The designer was free to make complete use of recent experience, to devise radical improvements."

The wind tunnels of Gottingen came to life, not to produce warplanes, but to refine a strange-looking craft that carried the wing on top and had no landing gear underneath. The fuselage was swept upward to support a high tail with no stabilizer, only a rudder and balanced elevator. In the *Vampyr*, the Wasserkuppe airmen saw the shape of the future—the sleek Hanover glider had increased the 5:1 glide ratio of the old Wright biplane to 16:1!

If the *Vampyr* proved to be a revolutionary design, a brand new flying technique was born at the Wasserkuppe a few days later when Frederic Harth, flying a homebuilt machine constructed with the help of the famous Willi Messerschmitt, soared for 21-½ minutes along the ridges near Hildenstein, following the graceful seabirds in sweeping figure eights, and finally landing less than 200 yards from his launch point. Harth had introduced the art of slope soaring, and the era of simple downwill skysliding was over.

For those who would pursue adventure in the high sky, the Wasserkuppe experiments of the 1920s would lead to a better understanding of how to harness the winds and thermals for both lift and propulsion in slender sailplanes that had to be launched with shock cords, winch, or auto or plane tow. The MPA spinoff would be the secondary serendipity.

On one unforgettable afternoon, when a towering, anvil-shaped cloud boiled and throbbed above the Wasserkuppe, a group of sailplane pilots launched into the wild sky to probe the storm's interior on a daring research project. One by one they were sucked inside the seething vapors to vanish in a churning blackness, where energies greater than that of a hundred atomic bombs were being unleashed by nature.

Stabbing lightning glowed eerily around the sailplanes and the heavens thundered. Violent updrafts hurtled them higher and higher, through stinging raindrops and pelting hailstones.

The pilots suffered from anoxia; their controls stiffened with ice. Most of the pilots managed to escape the raging thunderhead, but in his anger, Zeus, the thunder god, exacted a terrible toll for this invasion of his domain—by snapping the wings from three sailplanes. The birdmen, Lemm, Schultz, and Bleh, kicked free from their wrecked gliders and dove into the darkness, tugging on their parachute ripcords. For nearly an hour they rose and fell helplessly on the winds, inside the electrically-charged arena of death, then one by one dropped finally to earth. Their bodies were found later—one was charred by lightning, the others had frozen to death.

Said Professor Walter Georgii, director of the Rhon Rossiten Gesellschaft Research Institut at the Wasserkuppe: "Such experiences will give a new generation of flying men a body of weather wisdom by which they may safely meet and even turn to useful purpose the atmospheric disturbances so frequently met in air transport today. The true meaning of 'air sense' lies in this conquest of the variable atmosphere by the soaring pilot."

Atop the Wasserkuppe today stands a rocky cairn topped by a bronze condor, intently facing the winds from the west. The monument is a tribute to the heroic birdmen of Hesse, who gave their lives probing the unknown frontiers of the sky.

There is a close link between pure soaring craft and MPA's—both must be built with a sufficiently small wing loading to fly, the former on nature's updrafts, and the latter, in still air, on muscle power. Later on we will discuss the future of MPA's as a sport, which surely will result in combination machines wherein man-power can be utilized to stretch the glide of soaring craft when the thermals run out.

Thus in the 1920s, two different types of flying machines developed in Germany, where aerial engines were outlawed by the Treaty of Versailles—the true soarers, and the MPA's. The Peugeot Aviette activities in France only barely preceded a similar competition at Rhon, Germany, in the mid-1920s.

One interesting MPA developed at this time was a craft with odd rotating airfoils designed by a Silesian named Porella in 1925. Reay describes the machine as resembling a ship with paddlewheels, which produced both lift and thrust on their forward and downstrokes. A fixed wing of some 30 feet span and a conventional empennage were added. The pilot lay in a prone position and drove the paddlewheels with his feet. There is no record that it ever flew.

About 1935, Oskar Ursinus formed an MPA study group known as the *Muskelflug-Institut* at Frankfurt to assist independent researchers in the field with basic studies of such problems as power-train links with rotating (pedaling), reciprocating, and rudder-flutter motions. Herr Gropp, a young athlete, was the principal test pilot.

Already Ursinus's magazine *Flugsport* had posted a prize of 500 marks for the first MPA flight of one kilometer around two pylons 400 meters apart. No one claimed the prize, but a consolation prize

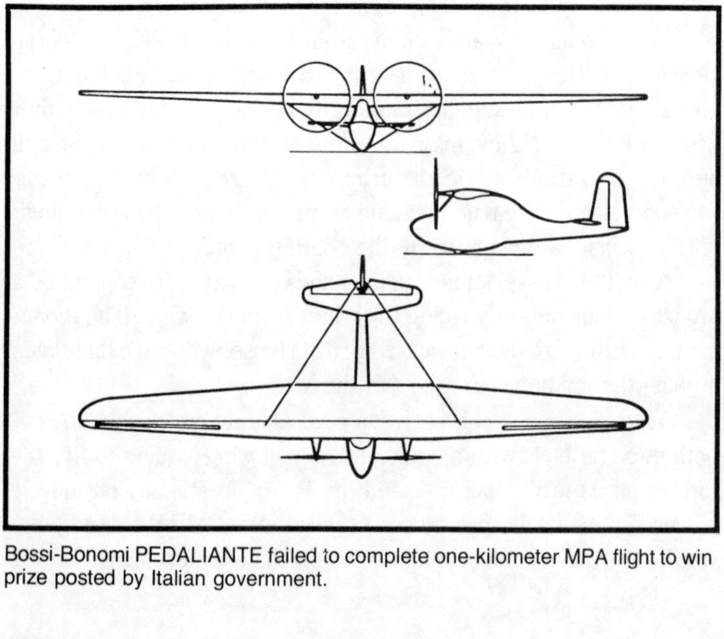

Bossi-Bonomi PEDALIANTE failed to complete one-kilometer MPA flight to win prize posted by Italian government.

went to a pair of Junkers engineers, Helmut Haessler and Franz Villinger.

The Haessler-Villinger machine, called the *Mufli*, was a single-seat monoplane resembling a typical sailplane, with a wingspan of 13.5 meters. The pilot flew in a reclining position, pedalling a foot-crank geared to a propeller, of 1.5 meters diameter, mounted on a pylon, by means of a twisted belt drive.

Mufli's cockpit was exceedingly cramped, and the pilot assumed an almost fetal position, clutching a joystick for lateral and pitch control. The wingspan was 44 feet, with a thick Gottingen 535 airfoil. Launched by shock cord, the machine once traveled 790 yards to win the *Flugsport* consolation prize.

In Italy in 1936, an Italian-born American, Enea Bossi, designed a glider-like MPA with two contra-rotating propellers to compete for a prize posted by the Italian government for the first MPA flight of one kilometer.

Bossi, following preliminary tests with various machines in America and in Italy, contracted with Vittorio Bonomi to construct a competition machine with a wingspan of 58 feet and a low aspect ratio of 13.4. Unfortunately, the Italian government insisted on their meeting rigid airworthiness standards, which resulted in an excessively heavy craft, estimated by D. A. Reay at nearly 100 kg.

Bossi favored large-diameter propellers and tried unsuccessfully to utilize an energy-storage device, consisting of twisted lengths of rubber shock cord. The pilot, Emilio Casco, flew the machine, called the Pedaliante, on numerous flights of considerable distance. The longest, according to Keith Sherwin, was a distance of 980 yards, from a shock-cord launch, mostly downhill.

Chapter 11

The Kremer Prize

Wars have a strange way of changing the course of events in unexpected directions. At the close of World War I Germany's aerial might was emasculated by the Treaty of Versailles, yet that instrument was responsible for the rise of technological development of both soaring craft and man-powered airplanes.

World War II, which cut off the supply of rubber from the Dutch East Indies, opened the way to development of synthetic materials and the giant petro-plastic industry. This, together with new breakthroughs in aerodynamic designs, finally provided the technology for development of ultralight flying machines that would have been impossible to construct earlier.

There remains one forgotten page of history in the man-powered aircraft story that was interrupted by the second World War—the *Colditz Cock*. In the waning months of the war, a group of British war prisoners, confined to Colditz Castle in Germany, resorted to a desperate means of escape, using an engineless glider with stored energy.

Working in shifts, four hours a day for ten months, they constructed a two-place "escape" machine in a concealed attic workshop. There they secretly fashioned a craft they nicknamed the Colditz Cock, from any materials they could scrounge. Wings were covered with bed sheets, doped with boiled milk; control surfaces were attached with cupboard hinges. A 60-foot catapult launch track made from mess tables was to straddle the ridge of the tiled roof, and a two-ton bag of rocks, tied to a rope, was to be dropped to provide launch power. Amazed Yankee invasion troops were shown the strange bird by the proud POW's, who never got the chance to see if she really could fly. The story came to light after the war in an article in *Air Review*, authored by the project's leader, Flight Lieutenant L. J. E. Finch.

In 1948 man-powered flight was discussed in the periodical *Aeronautics* by Brian Worley, who stressed the importance of minimum weight, and of low induced drag to be achieved by using a high aspect ratio wing.

Shortly thereafter the center of MPA activity shifted to the United States, where August Raspet, a noted aeronautical engineer at Mississippi State College, applied to the problem his extensive research work on sailplane design. Like Worley, Dr. Raspet recognized that flying close to the ground offered an opportunity to reduce the induced drag of the wing by some 50 percent. He further studied the laminar flow characteristics of birds' wings and applied that to a boundary layer control design, for adaptation to MPA's flying at low Reynolds Numbers.

Also interested in boundary layer control by means of wing suction devices was T. R. F. Nonweiler at Glasgow University, who in 1956 published his views on the potentials of man-powered flight. A key point was his careful determination of power required, through wind tunnel studies of the drag produced by a racing cyclist suspended inside the tunnel. He concluded that sustained power output of 0.40 kW was a reasonable figure for a cyclist in good condition.

About this time an unusual inflatable MPA was built and flown by Dan Perkins at the Royal Aircraft Establishment at Cardington, England, as a hobby. The delta-shaped single wing had a span of 27 feet and area of 250 square feet, and weighed 38 pounds. Dubbed the *Reluctant Phoenix*, it was made from one-ounce plain-weave nylon, proofed with 2 oz. of polyurethane for a total weight of 3 oz/yard.

The craft could be rolled up for easy transport, and in the field was inflated with a vacuum cleaner. Although its wing loading was suitably low, its low aspect ratio of 4.4 gave it a very high induced drag. Perkins made short flights a few inches off the floor of the Cardington airship hangar. He sat like a cyclist and drove the pusher propeller via a series of pulleys and a polyurethane-covered rope belt.

Considerably more interest in man-powered flight was generated at this time by the lectures and writings of Beverly B. Shenstone, an MPA enthusiast and chief engineer of British European Airways. To Shenstone, the problem was age-old—the ridicule heaped on MPA inventors made the subject difficult to be taken seriously.

What Shenstone proposed was to start from scratch and first establish just what the limitations of human power available were; second, draw a set of power-required and power-available curves for flights of required duration; and third, design the MPA to fit the curves. It was his belief that a tandem two-seater would work best, as it would have more than twice the power available—the second man could devote full energy to propulsion, while the first divided his energies between propulsion and control.

Another consideration was the relative inefficiency of the propeller (80%) and drive chain (90%), giving an overall efficiency of 72%. The wing design he said, should havc as much laminar flow as possible at a requisite Reynolds Number, on the order of 0.7×10^6, and have the bottom of the laminac flow bucket at a lift coefficient of the order of 0.9. Two possible airfoils, he suggested, were the NACA 65A (10)12 and the German Section FX05-H-126.

Unlike the Italian Government, the British Air Registration Board was more lenient on MPA ultimate load factors required for safety—a ULF of 2.5 would do well, thank you. That meant the design could have a quite flexible wing to save weight, while still avoiding flutter or control reversal at operating speeds.

Shenstone felt that the power train was the biggest problem to be solved—the best route to go, he suggested, would be to use pedals, sprockets, and chains, and a propeller of at least 6 feet diameter, possibly of balsa wood. As to overall design, he felt the tail-first (canard) type was more attractive than tailless or conventional types "because it is an easy job from the aerodynamic and control point of view and fits in well mechanically."

Bit by bit, Shenstone was outlining the basic requirements for a successful MPA that would become an integral part of the Kremer Competition regulations. "At present," he wrote in 1959, "there is no liklihood of the man-powered aircraft rising more than a few feet above the ground, because the power required within the ground cushion is so much less than the power required at about 15 or 20 feet. In fact, at about 10 feet altitude the ground effect doubles the effective aspect ratio of the wing, which has a large effect on the power required, because the wing incidence is quite high."

Shenstone concluded that "real interest in this problem is increasing because it is gradually being realized that the problem is one which is very close to solution, and the technical background is almost all available."

On January 10, 1957, a meeting of interested parties was held at the College of Aeronautics at Oranfield, at which was formed a Man-Powered Aircraft Committee (MAPAC) chaired by H. B. Irving. Committee members came from universities, the aerospace industry, and other groups such as the Royal Aeronautical Society and the British Gliding Association. MAPAC subsequently became a part of the Royal Aeronautical Society.

Without funds, MAPAC was kept busy making available MPA information to interested parties, but in November, 1959 their enthusiasm was backed by an offer of £5,000 from Henry Kremer, a British industrialist, for the first MPA to fly around a course specified by the Sociey's Group in consultation with the Royal Aero Club.

The Kremer Competition was laid out as a figure-eight course around two markers half a mile apart. Power for takeoff and during flight had to be muscle-power alone, without use of stored energy, or lighter-than-air gases. The aircraft had to cross the start and finish lines at a height of at least ten feet.

Limited to citizens of the British Commonwealth, the Kremer Competition had no takers, hence in March, 1967 Kremer added another £10,000 in prize money and the affair became international. As the goal seemed almost unattainable, early in 1973 Kremer raised the ante to £50,000.

In addition to the prize money posted, a fund was formed to assist in the development of the more promising designs, which appeared to be attainable only through costly research and development by team effort. In 1960 Air Commodore J. G. Weir donated £1,000. Kremer added to that, as did three leading UK aircraft firms, bringing the fund total to £5,000.

In March, 1967 another contest, the Kremer £5,000 Competition, was devised, consisting of two slalom runs in opposite direc-

British industrialist and health enthusiast Henry Kremer posted huge prizes for MPA competitive flights.

tions around three markers, and in 1976 Kremer also posted a prize of £1,000 for the first UK three-minute flight. Finally, there was the Robert Graham Undergraduate Project Competition, with an annual prize of £200, on any subject appropriate to man-powered flight.

Impetus of the Kremer Competition gave tremendous push to the MPA efforts around the world, and in England in November, 1961, two machines actually did achieve flight solely by their pilots' physical efforts. One was *Sumpac*, designed and built by research students at Southampton University, and now on permanent display in the Shuttleworth Collection. The other was *Puffin*, a product of the Hatfield Club at Hawker Siddeley's Airfield, which flew for 993 yards in 1962. *Puffin* was later redesigned and rebuilt at Liverpool University, where it flew briefly as the *Liverpuffin*.

PUFFIN Mark I flew 993 yards in 1962.

By 1970, a second generation of man-powered aircraft was beginning to appear. In November, 1971 the Weybridge Group's machine *Dumbo* flew, and in December 1972 the Hertfordshire Pedal Aeronauts' *Toucan* became the world's first two-man-

PUFFIN Mark I had pusher propeller behind tail.

The 1930's saw rise of "Batmen" like Jimmie Goodwin who "flew" during parachute freefalls.

powered aircraft to fly. Then in 1972 Lt. John Potter flew another British MPA, the *Jupiter*, for a distance of 1171 yards for an observed world record.

Potter's record was surpassed in January, 1977 by a machine built by students of Professor Hidemasa Kimura of Nihon University in Japan, with a flight of 2290 yards in 4-½ minutes.

These competition machines and others, described in the following chapters, all contributed something to the state of the art of flying by muscle power, culminating with the amazing flights of Dr. Paul MacCready's *Gossamer Condor* in California in August 1977, to capture the big Kremer Award. Beyond that, an exciting future awaits designers of both competition machines and MPA's to be flown purely for sport.

The *Gossamer Condor's* great flight prompted Kremer to post a still more generous award—a £100,000 prize for the first MPA to

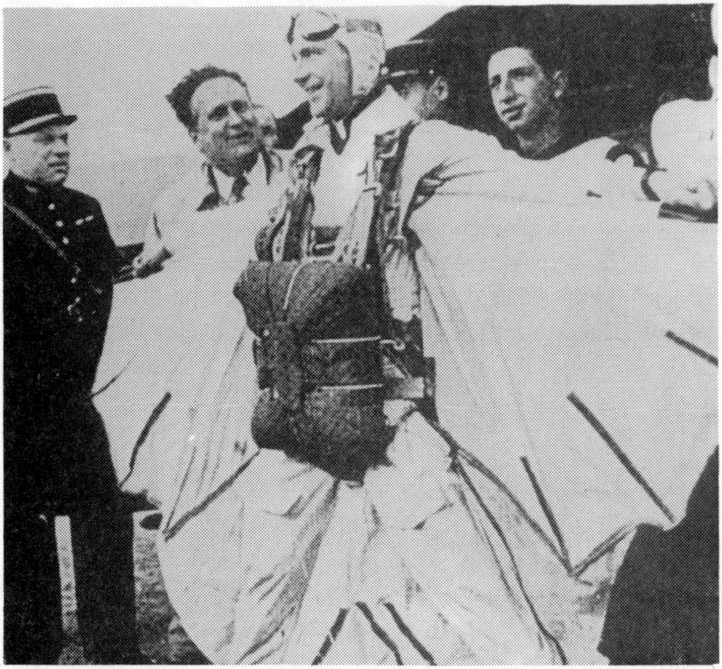

Batman Clem Sohn was killed in jump at Vincennes, France, in 1937 when his parachute cords became tangled with fabric wings.

Leo Valentin in freefall with wooden wings.

cross the English Channel, from England to France. The Kremer International Competition prize of £10,000 for the first figure-eight course MPA flight was limited to non-US citizens. And the Kremer Double Slalom prize money of £5,000 was limited to the first three British Commonwealth entrants to complete the double slalom around three markers.

Other Man-Powered aircraft grants, it was announced, would be available, at the discretion of the Royal Aeronautical Socity, MPA Group, to individuals or organizations who submit promising projects designed to further the achievement of man-powered flight.

Chapter 12
The Kremer Race Begins

British sporting blood was stirred by Henry Kremer's challenge. All over the United Kingdom aeronautical talk turned to the problem of man-powered flight, and the chances of winning the Kremer Award were hotly debated. Two MPA groups were immediately formed, each receiving a £1,500 grant from the MPA Group of the R. Ae. S.

Of these, one had the technical and financial backing of a giant of England's aerospace industry, the de Havilland Aircraft Company at Hatfield. The other consisted of a number of graduate aeronautical engineering students at the University of Southampton.

Where the Hatfield group had at their disposal de Havilland's giant wind tunnels and other facilities, the Southampton group was on its own. They had alrady begun MPA studies as a sort of hobby when the Kremer Award was posted. Their first step was to gather data on power output from all available sources.

A leader of the student group was a female—Anne J. W. Marsden, whose enthusiasm and inspiration gave impetus to their difficult research task. Writing in *Sport Aviation*, she recalls:

"We had teams of undergraduates running up and down stairs while we timed them with stopwatches, and finally we made a rig to simulate what we thought was a favorable position for the pilot, and measured the power output of a number of cyclists pedalling in the reclining position. The power was absorbed by electrical resistances from a pedal-driven generator and measured with a wattmeter.

"As a result of this work it was found that an average cyclist in the reclining position could provide about ½ hp for nearly ten minutes with his hands completely free to work any controls. This was made the basic of all our subsequent design work, although we allowed some good margins for error."

Head of the Hatfield MPA Club was a talented aerodynamicist, John Wimpenny, who in 1946 had already built a model MPA, which formed the basis for the "HMPAC" design.

Both MPA groups were limited in their design work by lack of availability of airfoil sections designed for performance at the low Reynolds Numbers appropriate to manpowered flight. Compared to current airfoil designs with lift coefficients of 1.2, with correspondingly low airfoil profile drag, the C_l values in 1961 were on the order of 0.85 at Southampton and 0.8 at Hatfield.

Thus the trend was toward conservative wing design with emphasis on low weight values. The Southampton machine, called *Sumpac* (Southampton University Man Powered Aircraft), had a span of 80 feet and a wing area of 300 sq/ft., and the Hatfield *Puffin I* a span of 84 feet and 330 sq/ft. of wing area.

Two other major differences were apparent—*Sumpac* carried its propeller mounted on a pylon, to avoid the long driveshaft required if the prop were carried at the tail, and the limiting effect on size by being lower to the ground.

The *Puffin* builders, on the other hand, favored the rear-mounted propeller to avoid the extra drag of the pylon. A balsa wood drive shaft broke, and was replaced by a metal tube. The entire transmission and drive linkage, which powered both the propeller and ground wheel, was developed by engineers of the Dunlop Company's Aviation Division.

For pilot training, the Hatfield group developed a flight simulator to solve a sticky problem—recovering from a bank was so sluggish that large ailerons and aileron deflections were found necessary, thus producing unwanted large aileron drag and yaw forces.

Meanwhile back at the *Sumpac* group, design limits were placed at 80 feet of wingspan, says Marsden, "because we didn't think our pilot would be able to manage to handle much more", flying in ground effect to cut induced drag. Further, they avoided assiduously using flying wires, struts, and exposed undercarriage, calling for a fully cantilevered monoplane wing, similar to the *Puffin's*.

Southampton University's SUMPAC had 80-foot wingspan, with inset ailerons and washout wingtips. Propeller was mounted on pylon to avoid long driveshaft to tail.

Wooden construction was decided on, with doped parachute nylon for covering. The *Sumpac* airfoil was the NACA 65_3818, a heavily cambered section which depends on maintaining laminar flow back to 60 or 70 percent chord. Tested at Southampton University's wind tunnel, it was modified by trial and error. *Sumpac's* biggest aerodynamic design problem was discovered in tunnel studies of a

1/6-scale model of the complete machine, at full scale Reynolds Numbers. At the root of the pylon, interference drag caused by breakaway flow was unacceptably high. A unique fairing was designed to correct the matter.

The propeller was shaped with built-up ribs over tubular metal spars, designed to produce 7 lb. thrust at 20 mph cruise speed. Tested in a slow-sped wind tunnel, it achieved 89 percent efficiency by using a turbulator strip glued to the top surface of the blades.

The propeller was shaped with built-up ribs over tubular metal spars, designed to produce 7 lb. thrust at 20 mph cruise speed. Tested in a slow-speed wind tunnel, it achieved 89 percent efficiency by using a turbulator strip glued to the top surface of the blades.

To get an elliptical spanwise loading on a tapered wing, severe washout of ten degrees was used, achieved by a twist of the double spars. The pilot sat inside a tubular framework, which supported the pylon and landing gear. A thin steel belt drive from the rear wheel to the propeller, twisted through 90 degrees, was found to slip badly, so a compound called *Belt Stick* was applied.

For a pilot, the *Sumpac* students wanted to use a well-known long-distance runner, Martin Hyman, but as he was not a pilot they substituted Derek Piggott, an experienced sailplane pilot. Piggott suggested that the first trials be held at Lasham Airfield in Hampshire, a glider port.

On a number of trial runs, Piggott found the control linkup confusing as it required "bobsledding"—use of right rudder to turn left. The linkage subsequently was reversed to standard aircraft practice. The first few trials were small disasters, with many breakages. The machine was stored in a war surplus Nissen hut that leaked badly, hence rerigging was required for each flight. By this time the morning winds were up.

Piggott also found the rudder area too small, so its size was doubled, but even then pedaling inside the enclosed cockpit was hard work, producing half a horsepower. The uniform of the day was shorts and singlet.

Rumors came that *Puffin I* was ready to fly, and had a far better chance of approaching the Kremer course rules than *Sumpac*—they had a closely guarded secret; their machine was covered with transparent Melinex plastic film, that could be shrunk to a smooth cover with a hot iron. The entire covering weighed only 2 pounds. Further,

PUFFIN II, redesigned from PUFFIN I, had 90-foot span with stong dihedral. Only fuselage was used from PUFFIN I in redesign.

their pilots had been rigorously training on a tight schedule, fed on Spencer steaks!

Piggott recalls the day when all was ready with *Sumpac*, and *Puffin I* had yet to fly:

"It was late in the evening of 9 November 1961, one of the most thrilling moments of my life. We had asked a cameraman to record what happened, we were that confident. It was damp and cold when I climbed into the seat and strapped myself in. The nose was bolted on...there was scarcely a breath of wind as I accelerated slowly down the main runway trying to keep straight, and watching the green pellet of the Cosim variometer, my airspeed indicator, which was slowly rising to the 18 feet per second mark at which I could ease back, and perhaps fly.

"Pedalling the machine was rather like riding a tandem bicycle up a gradual slope with the girl friend not bothering to pedal. Then I eased back on the stick...the pedals began to slip a little and it all became easier. Tremendous excitement! I pedalled harder trying to make it stay up for as long as possible.

"One wing began to drop and we turned slowly off the line of the runway. My reactions were slow, as I thought out the required aileron and rudder movements to stop the turn. Gradually she responded and the wings came level again. The elevator control was very sensitive and the aircraft pitched up and down as I overcontrolled. This caused extra drag and suddenly there was the rumble of the wheel on the grass. I stopped pedalling and concentrated on keeping it straight as I slowed down. I was puffed-out with trying so hard to keep up, but by no means exhausted. Anne rushed up to congratulate me, and my reaction was to congratulate her. Their machine had really flown at last, and we had film to prove it!"

Less than a week later, on November 15, 1961, the *Puffin I* was rolled out of the hangar at 10:15 p.m. and Wimpenney made the first taxi runs. The next day another pilot, Jimmy Phillips, made the first flight down the main Hatfield runway. Within another two days flights of up to one minute duration were accomplished.

Minor problems were discovered and corrected with *Puffin I*—oxygen deficiency was caused by a too-small air inlet in the cockpit area...forward vision was inadequate on hazy days, and in one crash, blamed on the lost horizon, half a wing was broken off.

A bike racer named Chris Church made several flights until one day early in April, 1963, a wind shear blew the machine off the runway. It crashed and collapsed in a heap of splinters. Rather than being discouraged, the Hatfield group took the opportunity to completely redesign *Puffin I* into *Puffin II*, which featured a wider wingspan of 93 feet and 60 square feet extra wing area, with 6 pounds more empty weight. From August 27, 1965, until retired two years later, *Puffin II* made several flights of more than half a mile. Its pilots learned something important on these flights—rudder alone served in making small turns or heading corrections. Perhaps ailerons were unessential after all!

Sumpac also made many good flights, one of more than 1,800 feet, and following graduation of the student-designers, it was moved to London by one of the group. It finally crashed for the last time on November 12, 1965, and was retired to the Shuttleworth Aircraft Museum at Old Warden Field near Bedford.

During their brief careers, both *Sumpac* and *Puffins I & II* were frequently flown with auxilary power supplied by model aircraft engines, to give the pilots more time to practice turns. Straightaway MPA flight was a proven success, but the Kremer course around a figure-8 had not even been attempted.

Although *Sumpac* and the *Puffins* were the first British MPA's to take advantage of the funds made available by the Man-Powered Aircraft Group of the RAeS, four other fixed-wing MPA projects were similarly funded, with grants ranging from £500 to £1,500. Of these, the Southend *Mayfly* two-place MPA never got off the ground. Built of spruce spars, balsa ribs and modellers' tissue, it

Southend MAYFLY never flew—it was damaged in hangar collapse.

encountered a series of mishaps, including a hangar collapse that did it in.

The single-place Woodford MPA, later to become the *Jupiter*, was nearly completed by its designer, S. Hodges-Roper, when it was badly damaged in a fire, but it was later rebuilt in the early 1970s to become a highly successful Kremer contender. The other two funded MPA'a were the Hertfordshire Pedal Aeronauts' *Toucan* and the Weybridge MPA.

Both the latter craft benefitted largely from lessons learned from the earlier Southampton and Hatfield efforts, according to John H. McMasters, an MPA designer and historian. The *Toucan* (Two Can Fly if One Cannot) had an impressive wing span of 123 feet and 600 sq/ft of wing area. It was designed by a group of employees of the Handley-Page Company led by Martyn Pressnell.

An interesting feature of *Toucan* was its midwing design, intended to make the most of ground effect, together with slotted ailerons and wingtip spoilers for faster control response. First flight took place in December 1972. Its longest flight was more than 1,000 feet.

The Weybridge MPA was constructed by a group headed by P. K. Green and included a number of employees of the British Aircraft Coporation. Like the *Toucan* it had a huge wingspan, of 120.3 feet, its extreme aspect ratio meant to lower induced drag. It featured a low wing and pusher propeller at the tail.

Initially formed in 1967, the Weybridge MPA Group eventually nicknamed their machine *Dumbo*. Its light weight proved a problem in gusty air, and first flight date was postponed until September 18, 1971. After a few more flights *Dumbo* was moved to Cranwell, where Squadron Leader John Potter's MPA took over the project, renaming it the *Mercury*.

WEYBRIDGE MPA airborne for first flight, flown by Chris Lovell, September 18, 1971.

All these efforts were lumped by McMasters into an initial design effort, Phase I (1958-1964), in which the designers were strongly influenced by the early opinions of Shenstone and Nonweiler, which led to development of the Kremer rules. Limited data on low-Reynolds-Number aerodynamics and model-airplane type materials were used.

Dumbo and *Toucan*, however, were also to be identified in a Phase II effort (1965-1971) overlapping the earlier work, and including four design goals within the Kremer Competition Rules, says McMasters. These were:

1. Reduce speed and lower wing loading.
2. Extend wings to gain a better L/D in ground effect.
3. Cruise at conditions closer to minimum power required.
4. Stress ingenious structural design to hold the weight of larger wings in check.

In addition to *Dumbo* and *Toucan*, another two huge MPAs were developed—the French Hurel *Aviette* and the amazing seven-place Canadian *Cochkanoff*, which was never completed. *Aviette* did cover a distance of some 1,000 yards before it was broken in a crash.

The results of Phase II were disappointing, says McMasters, proving that "big is not necessarily beautiful." The success of any MPA in competing for the Kremer prize would follow from a fresh approach to the problem taken by Flt. Lieut. John Potter's group of Royal Air Force apprentices.

Among their goals was development of better stability and control characteristics of MPAs, and a flight program that involved "jump" takeoffs out of ground effect and climbs to altitudes of 20 to

TOUCAN's midwing design took advantage of ground effect.

TOUCAN used auxiliary model airplane engine under right wing during trial flights. It once flew 1000 feet.

30 feet. In *Jupiter*, built at Halton RAF College in 4,000 man-hours by some 100 persons, they had what looked like a real winner.

Jupiter, with a span of 80 feet using the NACA 653618 airfoil, made a flight in 1972 of 1.23 kilometers with Potter as pilot. Flying

TOUCAN in 1975 during modification.

French MPA AVIETTE had 138-foot span, flew at Le Bourget Airport. Aspect ratio was 30:1.

high, Potter felt more comfortable than flying in ground effect, but flying figure eights remained the elusive goal, and for five years the Kremer Award would be safe.

Two men in England had quite similar views on the subject of man-powered flight—Henry Kremer and Dr. Keith Sherwin, of the Liverpool University's Department of Engineering. Dr. Sherwin was primarily interested in basic research, as was Kremer, chairman of the engineering firm of White & Riches Ltd. (another interest of Kremer was weight-lifting, which no doubt contributed to his interest in MPA's).

In September, 1969, the remains of *Puffin II* were donated to Dr. Sherwin after it was demolished in a crash into airfield equipment on its final flight. Rather than rebuild it to compete for the Kremer Prize, Dr. Sherwin used it as an undergraduate project to teach engineering design.

Renamed *Liverpuffin*, the machine was rebuilt with definite goals in mind—it had to be easy to build, manageable on the ground and in flight, able to fly up to 300 yards at 3 to 5 feet altitude, and easily transportable. The smashed balsa ribs of *Puffin II* were replaced with styrofoam formed with a hot wire, the first MPA in the UK to use this modern material, common today in ultralight sport aircraft.

No lateral controls were provided; instead, a 20 degree dihedral was built into the 64-foot wing. Wing area was 305 sq/ft., all-up weight 300 lb., for a wing loading at gross weight of 1.10 lb/sq. ft. The airfoil chosen was the Wortmann FX-63137, same as in *Puffin II*.

TOUCAN on 700-yard flight in 1973.

Fourth MPA built by Paul Stephaan Masschelein and Eric Verstraete, flew 820 feet at Calaid-Dunkirk Airport July 24, 1974.

Halton RAF College's JUPITER MPA had 80-foot span, was flown out of ground effect by John Potter. In 1972 it covered a distance of 1.23 kilometers.

JUPITER MPA during press demonstration flight.

Liverpool University's LIVERPUFFIN was rebuilt from wreckage of PUFFIN II by students of Professor Keith Sherwin in effort to win Kremer Prize. Note foam leading edge at end of center section, where outer wing panels attach.

Liverpuffin made her first flight March 18, 1972, with Dr. Sherwin as pilot, covering 20 yards at an altitude of 9 inches above the runway at Woodvale Airfield, Formby, Lancashire, proving to his satisfaction that styrofoam construction and rudder-only control near the ground was workable.

Cutaway view of LIVERPUFFIN shows pilot position.

Chapter 13

Yankee Doodlers

As had been expected, the Kremer Prize attracted its share of MPA designers we may kindly call unconventional, both in the UK and abroad.

Among the early contestants was a would-be birdman aptly named Donald Partridge. At 6 o'clock one foggy morning Partridge, wearing only slim trunks and a sweater, poised atop Hammersmith Bridge in London, vigorously flapping a pair of homemade wings. He stepped off into space and flew 50 feet—straight down into the muddy Thames.

Emiel Hartmann, a Chelsea sculptor fascinated with bird flight, did better on November 2, 1959, when his ornithopter reached an altitude of 50 feet above the jet strip at Cranfield, in tow behind an auto.

Alan Stewart, a Sheffield engineer, built a flapper with a wingspan of 14 feet, to be powered with both arms and legs. Nowhere.

Nick Walton, 23, sprinted down Creech Barrow Hill near Taunton in the West Country, with mighty flaps of his 22-foot wings, but failed to become an eagle.

Bob Wilson, at 39 the father of eight children, had energy left to pump away on his gyro-bike, a sort of autogyro with wheels within wheels, until a gust of wind toppled him. "We shall carry on," he announced.

Emil Hartman's MPA was towed aloft 50 feet.

E. Winter was improving steadily on takeoffs with his Air Screw Propeller Bike, the Glastonbury Zodiac (he scorned wings).

In Italy, Vincent Carnevali, the Roman Fly-Man, fluttered along the runway at Erbe Airport every morning for weeks to prove that da Vinci was right after all; in vain.

In Moscow, Dmitri Vladimitovich Ilyin, a Soviet test pilot, used a 3-horsepower engine to supplement his muscles in flapping a pair of wings, with no success.

Across the Atlantic, Britishers looked with some amusment at "Colonials" in the United States, who felt that good old Yankee ingenuity could achieve what solid British logic had failed to accomplish, when Kremer doubled his prize to £10,000 and opened the competition to all comers on February 19, 1967.

Alexander Lippisch, who left Germany and the Wasserkuppe for America after World War II, held to the belief that MPA's should be flown purely as a sport, and not for prize money. He suggested a small single-seater of balsa wood with a delta-shaped wing, but it was never completed.

A dozen other American projects were begun as feasability studies, reported one investigator, Robert Graham. These included groups at Purdue University, Massachusetts Institute of Technology, Georgia Technical University, Michigan State College, the Bell Telephone Laboratories, Grumman Aircraft Co. and the Philco Ford Corporation.

MIT's BURD II MPA was built of balsa and Mylar, featured end plates to increase effective span. Topsail and canard elevator was included.

At Purdue, Curtis Cole and John McMasters became fascinated with the subject. McMasters saw a great future in MPAs, but more as a sport than as a Kremer contest affair.

Of more than ordinary interest was the work at MIT, where Professor Eugene E. Covert supervised a student MPA project from its inception in January, 1970, when it began as a study of the current state of the art. The study convinced the group that an entirely new type aircraft was needed to improve performance.

With an eye to capturing the Kremer Prize, they developed a two-man biplane-canard configuration which was tunnel-tested at full Reynolds Numbers. For control, spoilers were chosen for roll and yaw, with the canard providing pitch control. Originally the MIT group called themselves *Bumpas* (Biplane Ultralight Man-Powered Aircraft Systms) but later the project was renamed *Burd*.

After six months' study of low-speed aerodynamics, high-efficiency propulsion systems, and low-density materials, a decision was made to pick the crew size. Four legs were better than two for propulsion, they decided.

A pusher propeller was adopted to minimize prop-wash interference on the aircraft, while the canard surface provided an effective visual reference for pitch and bank. Effort on drag reduction was geared to improving induced drag control rather than streamlining. *Burd* I was designed to fly at the speed for least power required rather than best L/D, to be compatible with the pilot's ability to maintain a steady low-power output. Preliminary design called for a wing area of 625 sq/ft., an aspect ratio of 25:1, and a gross weight of 450 lb. for a wing loading of 1.38 lb/sq. ft.

The Wortmann FX 61-163 was finally chosen and the wing loading reduced to .66 lb/sq. ft. Wire bracing between the top and bottom wings was considered acceptable in view of the emphasis on reducing induced drag. Tip deflections of only 4 inches were achieved, compared to 12 feet in monoplane designs like the Weybridge machine.

Disadvantage of the biplane configuration was its low aspect ratio of 13.6, hence tip plates were added to increase the effective aspect ratio. The canard was all-moving for pitch control, and wing spoilers on the top surface of the lower wings gave good roll and yaw control at once. As Wilbur and Orville Wright learned in 1903, the canard stalls first and drops, preventing main wing stall, a design feature also in current use in Burt Rutan's stall-proof VariViggen canard design, a light sport plane. It also was found that tip vortices from the canard at high angles of attack blanketed the spoilers when placed behind the canard, hence they were repositioned further outboard.

While the wing spar was built of balsa wood, aluminum tubing was used for the fuselage truss system. The Schwinn Bicycle Company provided the design for the two-man cycling positions. Nylon skin was used for covering.

The first flight resulted in nearly complete structural collapse, and the machine was rebuilt as *Burd II*. The primary changes were internal, to reduce weight. An accident to *Burd II* in the late fall of 1977 put the craft back in the shop, and although the *Gossamer Condar* had already flown the Kremer course successfully, the students decided to go ahead, regardless.

James M. McAvoy of Georgia Technical University in 1962 came up with a novel departure in design, for a low-wing monoplane featuring a 10-foot diameter, four-bladed ducted-fan propeller at the rear of its triangular cross-section fuselage. Enclosing the propeller in a shroud was intended to increase the propeller efficiency by reducing blade-tip loss, and the shroud also served as a rudder and

Prof. Eugene E. Covert of MIT checks two-man BURD II MPA.

James P. McAvoy of Gerogia Technical University designed this MPA in 1962, with four-bladed ducted fan propeller.

tailplane. Elevators were mounted at the top and bottom of the annulus. Differential wing flaps were added to control banking. A single stick and hand lever in front of the open cockpit provided control input. Robert Ritchie, the pilot, sprinted the machine down the main runway at Fulton Airport—and crashed.

In Southern California, where Professor Twining's "Flip Flop" ornithopter had been ridiculed in 1910, students at the Northrop Institute of Technology (now Northrop University) in the early 1970s designed an unusual channel-wing MPA with two counter-rotating, four-bladed propellers turning inside shrouds, similar to the McAvoy design. The students—Thurman Elliott, Jeffrey Kildow, Malcolm Smith, and William Sutherlen—proposed to construct the wing with Kraft paper honeycomb, paper Nomex honeycomb, both hexagonal and expanded cell, and thin vinyl, covered with 0.00035 inch Melinex. Two pilots were to pedal their *Flycycle* in tandem.

Like others before them, the NIT students had been turned onto MPA design work by the Kremer Competition, and in January,

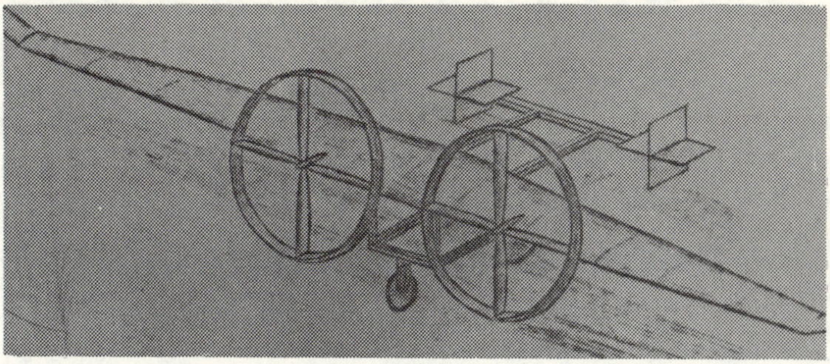

Early Northrop Institute of Technology MPA design featured counter-rotating propellers, twin tails, two-man propulsion.

1971, the project was transferred from Kildow's home to the NIT Aerodynamics Laboratory. A group of 20 other students joined them, with faculty guidance from Dr. Takashi Sugimura, assistant professor of aerospace engineering.

Eleven designs were tried before they settled on the final *Flycycle* design, a tandem two-place monoplane with a single pusher propeller. While the design work progressed, two two-man teams were chosen. Lead pilots were Steve Slaughter, a 17-year-old glider pilot, and Mrs. Rosemary Lichter, a sailplane pilot and wife of Lloyd Lichter, executive director of the Soaring Society of America.

Flycycle's wing spanned 80 feet on paper and would weigh only 15 pounds completed. Plans were made to flight test the NIT MPA at El Toro Marine Station, then move it to the Salton Sea area for a shot at the Kremer Prize in the dense, below-sea-level sky.

Later NIT MPA design was two-seater with pusher prop. It never flew.

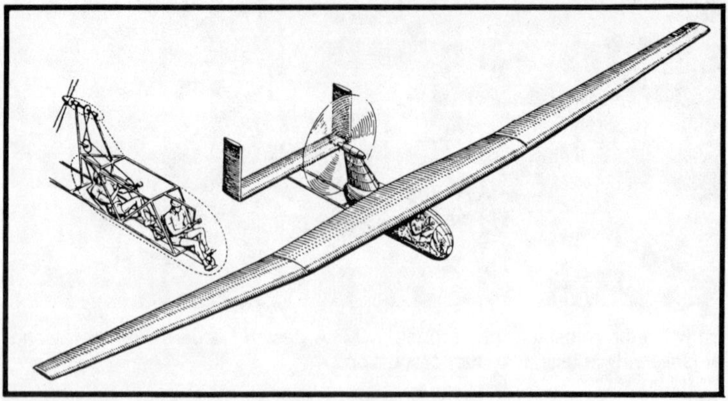

Melville M. Murrell switched MPA design to glider and made flights in 1877 at Panther Springs, Tenn.

Graduation of the Northrop students ended the project temporarily, but in 1973 it was briefly reviewed by a new student group working under Professor Johann Arbocz. A complete computer analysis of the Wortmann airfoil was run before interest lagged again. The project was never revitalized before the *Gossamer Condor* flights.

"Yankee Doodlers" in fact were busy long before the Kremer Competition got under way. Historically, one of the first American MPA's ever to get off the ground was Melville M. Murrell's *Aerial Navigator* of more than a century ago. In 1877 he patented an ornithopter in which the operator pushed on a crossbar with his feet to make the wings flap. The wings consisted of a number of hinged slats, normally kept closed by springs, which opened only on the up-stroke by air pressure from above. A pivoted tail was so rigged that a "wriggling or partially-rotary motion" could be imparted by the pilot.

Born in 1855, young Melville enjoyed tinkering with tools in the barn of his father's big home at Panther Springs, Tennessee, on the Knoxville-Bristol post road. Only 22 when he won his United States Patent, Melville altered his MPA design somewhat before flying it. The slats were mounted to provide foreward propulsion rather than lift, and the wing was a cloth-coverd rectangle. The pilot rod underneath the wing in a tubular framework. A cannard surface was mounted forward of the main plane.

In this machine the young inventor made several flights, launched at the end of a tow rope, and frequently covered several

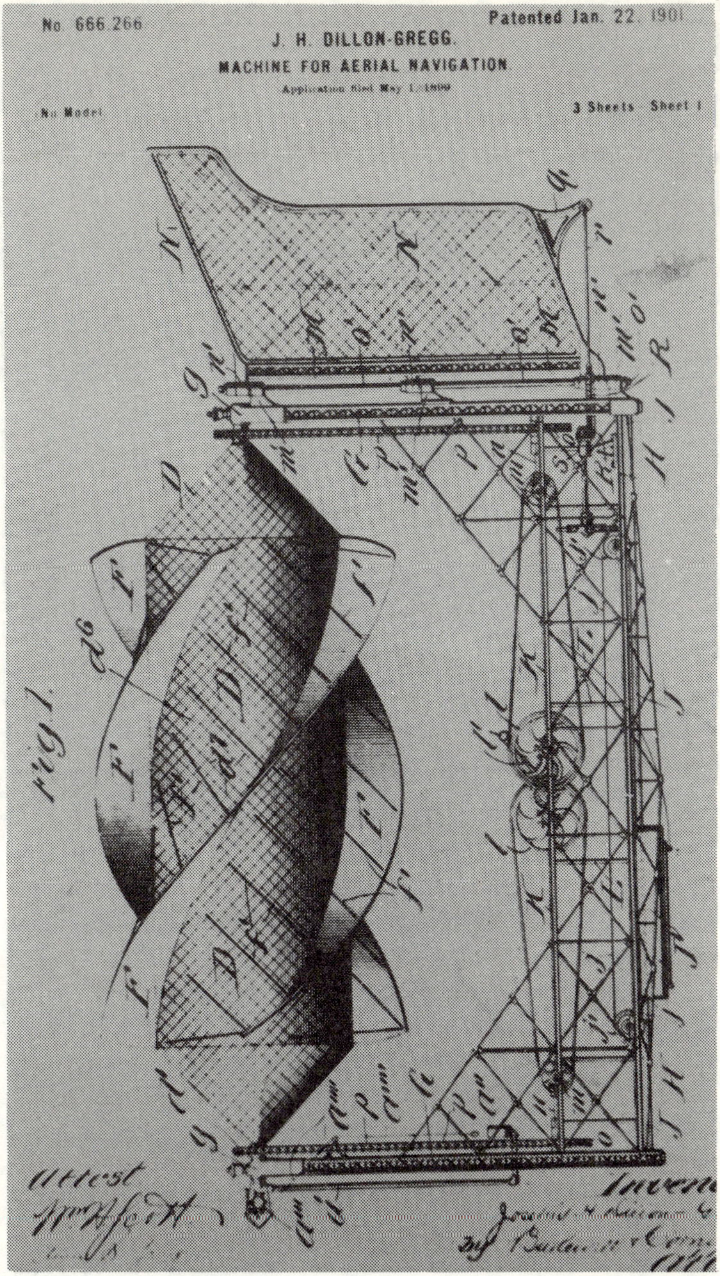

Archimedian screw principle was used in Jean Lassie's 1856 "Navigable Balloon" design, Craft was 900 feet long, designed to carry 300 men inside cylinder, with 150 men running on a treadmill to provide power. Trim weight is suspended below. (Hatfield Aeronautical Library)

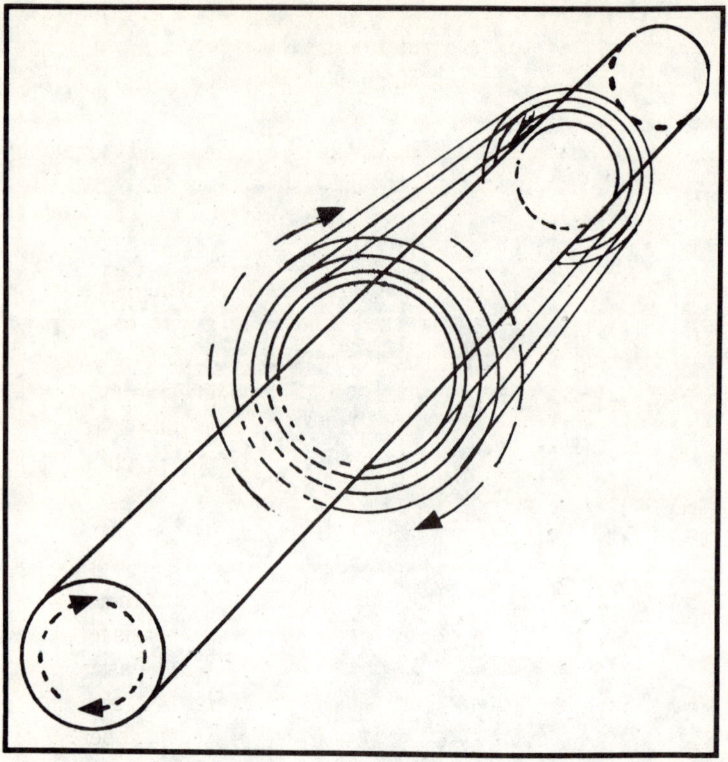

Magnus Effect.

hundred yards, according to family records. Historically they stand as the first recorded man-carrying glides in a heavier-than-air machine in the United States,and in fact preceded Lilienthal by more than a decade.

His goal was to achieve true powered flight, but lack of an adequate engine frustrated him, and soon he turned to religion, riding circuit as a backwoods preacher for the next 45 years. Many times the Reverend Murrell prayed that he might fly over Panther Springs in an aerial machine, his son recalls. In all, he did fly his device over the family apple orchard three times, no mean feat at the time. Murrell dies in 1932, satisfied to know that powered flight was no idle dream.

A quite different approach to man-powered flight has occupied the minds of researchers since the turn of the century—the Magnus Effect, in which a cylinder rotating around its spanwise axis in an air flow sets up a circulation and generates lift. D. A. Reay, a British engineer and author of *The History of Man-Powered Flight* (Pergamon Press, 1977), in 1965 undertook a research program at Bristol University in unconventional lifting-devices including Magnus Effect cylinders.

In 1969 in the United States, William F. Foshag and Gabriel D. Boehler of Aerophysics Company in Washington, D. C. published a paper on "HARWAS"—Horizontal-Axis Rotating-Wing Aeronautical Systems—listing some 1200 references to this subject, concluding that "promising new aeronautical systems" exist in this design.

Among the HARWAS devices tested was a wing rotor glider demonstrated by Koppen in 1903, and a Navy-Assisted Magnus Effect Aircraft by Ames in 1910. The Zaparka Magnus Rotor Aircraft possibly flew briefly in full size in 1930, they state, and an eight-foot wing rotor sailplane showed good flight performance and stability about all axes.

The Magnus Effect is basic to an understanding of the Kutta-Zhukovski theory of total circulation (see Chapter 4). It works this way: A cylinder spinning in a fluid at rest with respect to the cylinder sets up a circular and purely rotational flow, due to viscous forces in the fluid. If the cylinder is set in motion, the airflow assumes a streamline pattern with compression of the streamlines above the cylinder, a widening below it, and a downward component behind it, thus producing a vertical lifting force, or Magnus Effect.

An airfoil moving through still air produces a rotational flow in a similar manner to that around a cylinder, with a bound vortex around the wing and a starting vortex at the trailing edge, developing a downward momentum corresponding to the lift generated, as with the spinning cylinder.

Clyde D. (Daring) Goehring of Oceanside, Calif., experimented with MPA's at 72.

Third Archimedian screw propulsion idea for a MPA was used in the FLY WORM built in 1929 by Paul Maiwaru of San Diego, Calif.

Almost every day, new and unusual MPA's appeared in the public press, and too often were subjected to ridicule. Some progress was made, however, in various parts of the country. Wayne Bliesner of the University of Washington succeeded in making a towed flight in his winged machine before the hanger roof fell in and destroyed it. He was rebuilding it at last report.

In Oceanside, California, 72-year-old Clyde D. "Daring" Goehring in 1977 was at work on a pedal pusher tractor monoplane MPA with a huge propeller up front. Goehring drew on past experience as a wind tunnel technician at Douglas Aircraft Co., but he was still ground-bound when *Gossamer Condor* took off to make history.

And then there was Nick Dee of Los Angeles, who in 1972 was working on a secret MPA design that included a sort of slotted parachute that let him jump four feet off the garage floor and come down nice and easy. Later he changed the design so he could lie flat under the wings and drive a propeller with both hands and feet. It worked—too well. He flew smack into the wall and busted the prop.

Chapter 14
The Winner!

July, 1977, passed quickly at Shafter-Kern County Airport. The *Gossamer Condor* had by now made more than 430 flights since its conception just a year before. It had accumulated more flight time and traveled a greater distance than all previous MPA's combined, and yet the Kremer Award was still intact.

Time was running out. It was now or never, Dr. Paul MacCready realized. The secret was out—the world knew how the giant flying machine crept around corners, with its reverse-action wing-warping mechanism. In mid-August, *Gossamer Condor* was rebuilt for the last time, following the crash that occurred when a trailing vortex from a landing Ag Cat snapped its frail wing.

The fuselage was reworked and the crank mechanism lightened by drilling holes through the wheels, and now the craft stood seven pounds lighter. Jack Lambie estimated that meant a reduction in power-required by 0.2 horsepower.

Then came news from overseas—Professor Kimura's Stork was warming up for a final shot at the Kremer Prize! The chief pilot, 5′ 7″ T. Kato, had been in training by riding his bike 50 kilometers a day, and already had made a remarkable flight of 6,807 feet at Shimofusa Airfield; he might have done still better, had the runway been longer.

That flight was an official world record, observed by officials of the Japan Aeronautic Association as covering 2074.9 meters in 4 minutes 43 seconds, nearly doubling the former world distance MPA flight of 1071 meters held by John Potter in England. Dr. MacCready felt uneasy—would the Japanese learn his secret and be able to fly the figure-eight Kremer course?

When I visited Dr. MacCready earlier that year, covering the *Gossamer Condor* flights for Homebuilt Aircraft Magazine, I was sworn to secrecy about certain tehnical aspects of the design, and adhered to my promise. But now the full story can be told.

Professor Emeritus Hidemasa Kimura of Nihon University had encouraged his engineering students to develop a man-powered aircraft, soon after word arrived in 1961 of the flights of Britain's *Sumpac* and *Puffin* machines.

As a first step, in 1963 an experimental testing device was designed to measure the power exerted when a man pushes against a bicycle pedal. The second year was devoted to operations research into optimum airframe dimensions, weight, aerodynamics, and other factors. And in February 1966 Japan's first MPA rolled out—the *Linnet*, named for a tiny bird whose flight performance, in Professor Kimura's words, "is rather unimpressive."

Japanese MPA Linnet first flew in February, 1966.

He reported: "On February 25, 1966, our dear *Linnet*, with Munetaka Okamiya at the controls, lifted wheels off the gound for the first time. The following day the bird flew some 15 meters, marking Japan's first successful steady flight of a man-powered aircraft."

Because of his aerial feat, says the professor, "Mr. Okamiya became the first Japanese who was able to walk on the ground, swim at sea, and fly in the air using his own power."

In all, five *Linnets* were built, mainly as a research project for the students, not to compete for the Kremer Award. Secrets of the *Linnet* design were light-weight construction and use of styrene paper for covering, superior to the Melinex used in England. A student got the idea when he spotted some sashimi (raw fish) being sold wrapped in styrene.

A design drawback was the requirement for a long transmission shaft to turn the propeller at the tail, which resulted in considerable vibration. Pilots chosen to fly the *Linnets* were all licensed private pilots.

A new design series called *Egret* was launched in 1972, and the tail propeller position was abandoned for a pylon mount and a belt drive, similar to the British *Sumpac* and *Jupiter* machines. The rear fuselage was lengthened to provide greater longitudinal stability. Of three *Egrets* built and flown, the first was destroyed by turbulence, but the others flew quite well.

By 1975, the Nihon freshmen class had become seniors, with considerable design experience behind them. One student, Junji Ishii, was an expert designer, and Professor Kimura left him in charge of the new MPA, called the *Stork*.

Aerodynamically improved over *Egret*, the *Stork* featured detachable wings for easy transport. Its overall weight was amazingly low—36.0 kg—and yet it was quite rigid. Parts of the machine were covered with a Japanese paper called *ganpi-shi*. Every ounce of excess weight was removed, and a chain drive replaced the troublesome belt drive, linked to both the propeller and a ground wheel.

The initial *Stork*, called Model A, first flew on March 12, 1976 and in all made 45 flights including 11 over 400 meters (1312.33 feet). On June 4, 1976, unknown to the *Gossamer Condor* team in California, a student pilot, K. Churei, succeeded in making a 180-degree turn, at a 10-degree bank flying 9 feet above the ground.

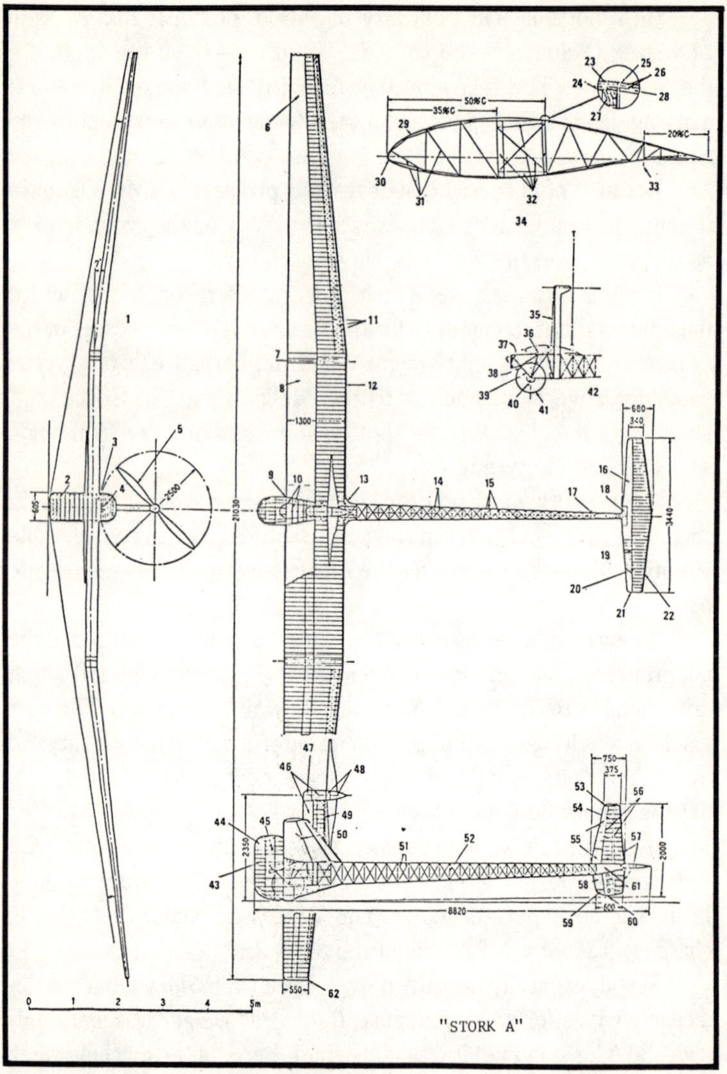

Three-view of Japanese MPA STORK A.

The turn wound up into a spiral dive, however, and the machine struck the ground. Professor Kimura was elated, nevertheless, and was convinced *Stork* could fly the figure-eight course at a smaller bank angle. *Stork*, he believed, had reached the level of the British MPA performances.

Rebuilt as *Stork B*, the Japanese MPA's wing was waxed to cut friction drag, the fuselage reinforced with balsa bulkheads, and the

Nihon U. student works on STORK's propulsion system.

cockpit canopy reshaped to fit their top pilot, T. Kato. The elevator movement also was decreased, against the amount of the twist of the control lever.

Officially designated the NM-75 *Stork*, the cantilever shoulder-wing monoplane used the Wortmann FX61-184 airfoil at the root, and FX-63-137 at the tips.

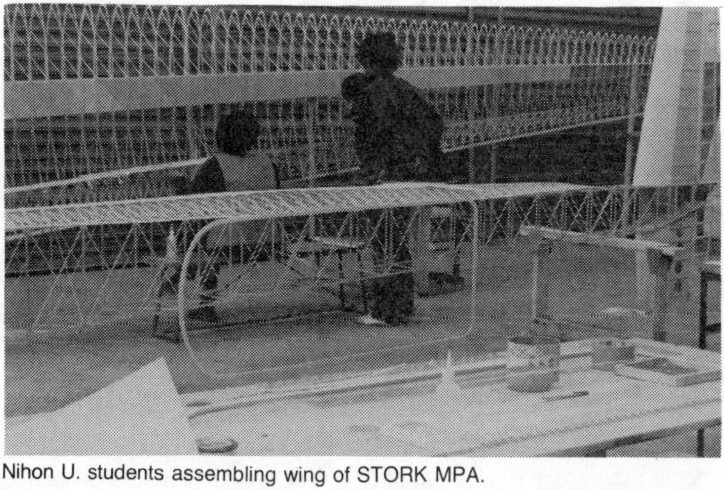

Nihon U. students assembling wing of STORK MPA.

Nihon U. students work on STORK's wing.

With a wingspan of 68 feet 11 inches and wing area of 233.6 sq/ft., STORK's wing loading was only 0.89 lb/sq. ft. Its optimum speed was 19.2 mph with a C_l of 0.94, and the power required in ground effect was computed at 0.33 hp, or 0.35 hp at 3 meters altitude.

Professor Kimura had other plans for *Stork* than simply shooting for the Kremer Award—it served well as a student engineering project, and plans were made to install solar panels along the wing's upper surface, to power an electric motor with solar energy.

Closer to home, Dr. MacCready kept a weather eye on conditions at El Mirage Dry Lake on the Mojave Desert of Southern California, where Taras Kiceniuk was flight-testing his *Icarus* ground effect MPA, designed to skid through its turns rather than bank.

Icarus' fuselage was built from aluminum and styrofoam, with an aluminum tail boom sticking out behind. The empennage was built up from foam sheeting, with the control cables running forward through the tail boom to the control handle-bars. The tail boom also served as the propeller shaft. The propeller blades had variable pitch, to maintain a constant rpm.

The wing construction was unusual, consisting of 56 individual styrofoam blocks cut out with a hot wire. Each block was hollowed for lightness, then all were glued together over the full 41-foot span.

Flight of STORK B at Shimofusa Naval Air Base, Japan.

MAN-POWERED AIRCRAFT OF NIHON

	AIRCRAFT	LINNET I	LINNET II	LINNET III
	DATE OF FIRST FLIGHT	26 Feb.'66	19 Feb.'67	26 Mar.'70
	MAXIMUM DISTANCE FLOWN (m)	43	91	31
DIMENSIONS (m)	WING SPAN (m)	22.30	22.30	25.30
	LENGTH O.A.(m)	5.60	5.80	5.86
	HEIGHT O.A.(m)	4.185	3.52	4.14
	TAIL PLANE SPAN (m)	5.40	5.40	5.80
AREAS (m^2)	WING, GROSS (m^2)	26.0	26.0	30.2
	AILERONS, TOTAL (m^2)	2.20	2.20	– (1)
	FIN (m^2)	0.97	0.97	1.53
	RUDDER (m^2)	0.46	0.46	0.45
	STABILIZER (m^2)	2.18	2.18	– (2)
	ELEVATORS (m^2)	1.46	1.46	3.40
WING	AEROFOIL SECTION	NACA63_3-1218	NACA63_3-1218	NACA8418-8415
	WING THICKNESS(%)	18.0	18.0	18.0~15.0
	ASPECT RATIO	18.5	18.5	21.2
	TAPER RATIO	3:1	3:1	3:1
	DIHEDRAL	3°	3°	6°
PROPELLER	DIAMETER (m)	2.70	2.70	3.00
	r.p.m.	160	200	125
	SPEED OF AIRCRAFT (m/s)	8.2	8.0	7.6
WEIGHTS	WEIGHT EMPTY (kg)	50.6	44.7	49.8
	FLYING WEIGHT (kg)	108.6 (3)	102.7 (3)	107.8 (3)
	WING LOADING (kg/m^2)	4.18	3.95	3.57
WEIGHT BREAKDOWN (kg)	WING	22.83	19.23	22.0
	TAIL UNIT	4.03	3.23	3.1
	CONTROL SYSTEM	0.86	0.96	1.0
	FUSELAGE	11.90	12.68	12.0
	UNDERCARRIAGE	2.55	2.55	3.5
	PROPELLER	} 4.78	1.03	1.3
	TRANSMISSION		3.66	5.5
	MISCELLANEOUS	3.65	1.36	1.4

(1) no aileron (2) flying tail (3) weight of pilot = 58 kg.

O.W. outer wing N.A. no data available

	LINNET IV	EGRET I	EGRET II	EGRET III	Stork A	Stork B
	13 Mar.'71 60	28 Feb.'73 34	30 Oct.'73 154	16 Nov.'74 203	12 Mar.'76 595	24 Nov.'76 2093.9
	25.30	22.70	22.70	23.00	21.00	
	5.80	7.40	7.71	7.70	8.85	
	4.14	2.29	3.66	3.70	2.40	
	5.91	4.00	5.00	4.70	3.44	
	30.2	28.5	28.5	28.5	21.7	
	— (1)	1.90	3.11	2.40	2.52	
	1.55	0.75	1.04	1.01	0.81	
	0.45	0.50	0.56	0.76	0.35	
	2.33	— (2)	1.98	2.13	— (2)	
	1.55	3.30	1.32	0.70	1.71	
	NACA8418	FX61-184	FX61-184	FX61-184	FX61-184 ~FX63-137	
	18.0	18.4	18.4	18.4	18.4~13.7	
	21.2	18.1	18.1	18.6	20.3	
	3:1	2.5:1 (O.W.)	2.5:1 (O.W.)	2:1 (O.W.)	2.36:1 (O.W.)	
	6°	6° (O.W.)	6° (O.W.)	6° (O.W.)	7° (O.W.)	
	3.00	2.70	2.70	2.70	2.50	
	125	180	180	120	210	
	7.7	8.5	9.6	9.8	8.6	
	55.0	57.0	55.7	61.1	35.9	
	113.0 (3)	115.0 (3)	113.7 (3)	119.1 (3)	93.9 (3)	
	3.74	4.04	3.99	4.18	4.33	
		25.3	24.0	33.0	19.9	
		3.0	3.0	2.8	1.0	
	N.A.	5.0	4.7	2.9	} 7.4	
		11.5	15.0	} 15.0		
		3.0	3.0		1.0	
		1.2	1.0	1.4	1.0	
		5.0	5.0	6.0	5.6	
		3.0	—	—	—	

Taras Kiceniuk (left) and assistants Bill Watson and Dave Sachs with ICARUS ground effect machine, canopy removed at El Mirage Dry Lake.

A small wheel was built into each wingtip, and the wing spar was carved from a piece of 1″ by ½″ Douglas fir, tapering to ⅛″ at the tips. Wings and fuselage were covered with a total of 60 rolls of red and blue translucent solar film, donated by a commercial firm, Pactra.

On August 1, 1977, Taras Kiceniuk took *Icarus* out to El Mirage Dry Lake for its initial flight test, but its high-aspect ratio wooden propeller stalled at about 1/3 its anticipated thrust. A foam propeller was substituted and worked better.

Three different pilots flew *Icarus*—Kiceniuk, Bill Watson, and Dave Sacks, the latter a glider pilot and bicyclist. On each flight the machine was towed aloft behind a small motorbike at about 20 mph, to let the pilot get the feel of the control system. As beautiful and graceful as she looked, skimming over the alkali lake bed, *Icarus* was still far from attaining anything like the ungainly *Gossamer Condor's* performance.

But there was still another MPA for Dr. MacCready to worry about—Joe Zinno's *Olympian* ZB-1, which became the first MPA to fly in the United States on April 21, 1976, from a runway at Quonset Point, North Kingstown, R. I. A retired Air Force Lieutenant Colonel, Joseph A. Zinno lifted off at 8:10 a.m. in a quartering wind of 8 knots, rose a foot, and covered 77 feet before touching down.

Zinno was tracked by Navy radar at 15.4 fps for 5 seconds, at a ground speed computed at 9.77 mph following a liftoff speed of 17.77 mph. Average ground speed was timed at 10.5 mph.

"The most important thing," Zinno told the author, "is to solve the problem, not win the Kremer Award, no matter how long it takes. The problem is definitely solvable, and with the growing number of people interested in MPA flight, each with his own approach, somewhere along the line, if we could find a clearing house for all these ideas, we might find the making of a solution."

Working alone with limited resources, Zinno, an ex-USAF pilot with 8,000 logged hours, had flown every type aircraft there was, "and it got so that one airplane was just like any other. There was no challenge left!" The next thing he knew, "There I was with this crazy model airplane on my hands!"

Bill Watson pumps air with cardboard box pump into ICARUS cockpit to keep it cool.

Taras Kiceniuk's ICARUS MPA flies in tow behind motorcycle.

Zinno's *Olympian* ZBl-1 was basically a midwing cantilever pusher with a 78-½ foot span, conventional tail and a flying stabilator attached to a 4-inch diameter aluminum boom. Propeller diameter was 8.7 feet, the wing built up around a box spar. Empty weight was 150 pounds, design gross weight 290 pounds. Wing area was 312 sq. ft., optimum loading under 1 lb/sq. ft.

Controlled conventionally from an enclosed cockpit, the Zinno MPA used bike-type hand grips, for rudder control on one side and for the elevator on the other. He used a treadle-type propeller drive system. He went to the Wortmann 150B airfoil near the wing root, with a tremendous camber to generate high lift at low speed, and a Wortmann 137 airfoil outboard "to compensate for a hellatious diving moment."

So that was the main known competition on August 23, 1977, when Bryan Allen eased into the cockpit of the 70-pound *Gossamer Condor*, at Shafter-Kern County Airport, for the 223rd time. All systems were go. Smoothly he accelerated into a shifting, 2-knot zephyr that drifted him slightly off course. He sailed easily over the ten-foot hurdle to enter the first turn to the right, around the first pylon.

Cries of encouragement spurred him on as he pedaled rhythmically toward the next pylon. Rounding that, to the left, he saw Jack Lambie holding the hurdle bar in the distance—the finish line! He later related:

"I was busy pedaling and flying the *Condor*, judging the turns to make sure I was outside the markers and correcting for the wind

Col. Joe Zinno, USAF (ret.) poses with his MPA, the Olympian ZB-1, that flew 77 feet.

which kept trying to move us off course. Then I saw the T bar ahead of me and suddenly, almost shocked, I realized that this flight would be it!

Prize-winning flight of GOSSAMER CONDOR on August 23, 1977, that won coveted Kremer Award of $50,000. (Photo by Judy MacCready).

Bryan Allen, who flew Gossamer Condor to victory.

"I speeded up my pedaling a bit and topped it with three feet to spare. I was tired but not exhausted so I slowed but flew another 400 feet before landing. Dr. MacCready came over, with a smile as wide as his face."

To Paul MacCready, success of the 7½-minute flight belonged to the full team, not to any one individual. His MPA had finally won the coveted Kremer Award, but the future lay ahead. What good was it? The *Gossamer Condor* had become a docile behemoth, so easy to fly that a number of men, women, and children who had never flown before, successfully flew it on short hops. He observed:

"The Kremer Prize flight represented an aviation milestone, in the sense of reaching one form of man's oldest aviation dream. Unlike some other aviation milestones, there does not seem to be any direct practical or commercial application which can follow.

"Human powered flight may help man's spirit, may raise his appreciation for his physical potential, and serve as an incentive for people on physical development programs, and may give him a better understanding of the flight challenges of soaring birds, but he will not pedal to the market or to work."

Teenager Monica Lake preflights Gossamer Condor.

Dr. Paul MacCready.

AWAY SHE GOES! Monica Lake flies the Gossamer Condor.

He did, however, see a potential in combining the concepts of MPA's and hang gliders, as well as other slow-flying ultralight vehicles, as an eventual basis for very light, inexpensive gliders of spans between 35 and 50 feet, able to take off from foot-launch, climb to altitudes using motors in the 2-to-6 horsepower range, and then outsoar the hawks. After all, it was that premise that first got him started thinking about MPA's.

Another more immediate goal had also caught Dr. MacCready's attention—the new £100,000 Kremer Cross-Channel Competition for the first MPA flight from England to France. With the $87,500 Kremer money pocketed, at this writing he had already begun initial design work on a Channel-Hopper MPA, one that would not require dangerous wave-skimming flight, in the manner of dynamic-soaring sea birds.

The shortest distance across the English Channel is roughly 22 miles, at the Straits of Dover, and Dr. MacCready sees no reason why, on a calm day, a redesigned *Gossamer Condor* cannot make it across.

Index

PICTURE SOURCES

AA -American Aeronaut, Sept., 1909.
AC -Astra Castra, Experiments and Adventures in the Atmosphere, by Hatton Turnor; Chapman & Hall, London 1865.
AIA -Airbus Industrie's "Aspects."
ANF -Artificial and Natural Flight, by Sir Hiram Maxim; Whittaker & Co., London/NY, 1908.
APS -American Philosophical Society.
BC -Bruce Carmichael.
BCC -Bureau of Commerce, Calais.
CA -Codex Atlanticus; Biblioteca Ambrosiana, Milan, Italy.
CDG -Clyde D. Goehring.
DD -Don Diggins collection.
EEC -Dr. Eugene E. Covert, MIT.
FAA -Federal Aviation Administration.
GC -Charles H. Gibbs-Smith, Sir George Cayley's Aeronautics 1796-1855; Her Majesty's Stationery Office, London, 1962.
HAGB-History of Aeronautics in Great Britain, by J. E. Hodgson; Oxford University Press, London, 1924.
HAL -Hatfield Aeronautical Library.
HDA -Historie de L'Aeronautique, edited by L'Illustration; Paris, 1932.
HK -Prof. Hidemasa Kimura, Nihon University.
HW -Harper's Weekly.
IFD -Institut de France.
JR -Joel Rieman photos.
JWRT-John W. R. Taylor.
JM -Judy MacCready.
JM -John Moody.
JZ -Joseph Zinno.
LACM-Los Angeles County Museum.
LAT -Los Angeles Times.
MDC -McDonnell Douglas Corp.
MJ -Mike Jerram photo.
NA -National Archives.
NIT -Northrop Institute of Technology.
PFM -Progress in Flying Machines, by Octave Chanute: The American Engineer & Railroad Journal, N.Y., 1894.
PO -United States Patent Office.
PSM -Popular Science Monthly.
PM -Popular Mechanics.
RAeS-Royal Aeronautical Society.
RGM -RG Moultou.
SA -Scientific American.
SDC -S. D. Coleman photos.
SI -Smithsonian Institution.
SM -Science Museum, London.
TIS -Travels In Space, by E. Seton Valentine and F. L. Tomlinson; Hurst & Blackett Ltd., London, 1902.